The Quest for Man

With a Preface by Sir Julian Huxley

VANNE GOODALL accompanied her daughter, Jane van Lawick-Goodall, to Gombe and helped to found the research center there. She is co-author with L. S. B. Leakey of *Unveiling Man's Origins.*

BARBARA BENDER is a Lecturer in the Anthropology Department of University College, London. She is the author of two books, *The Transition from Hunter-Gatherers to Farmers* and *Archaeological Guide to Normandy, Brittany, and the Channel Islands.*

THEODOSIUS DOBZHANSKY is Adjunct Professor of Genetics at the University of California, Davis. He is the author of many books, the best-known of which is *The Biology of Ultimate Concern.*

IRENÄUS EIBL-EIBESFELDT is a professor at the University of Munich and Head of the Research Unit for Human Ethology of the Max-Planck Institute. He has conducted many research expeditions, and he is the author of several books and papers, including *Land of a Thousand Atolls* and *Love and Hate.*

DAVID A. HAMBURG is Professor of Psychiatry at Stanford University School of Medicine, California. He is a regular contributor to learned journals and the author of *Psychiatry as a Behavioural Science.*

JANE VAN LAWICK-GOODALL, founder and Scientific Director of the Gombe Stream Research Centre, is also an associate visiting professor at Stanford University in human biology and psychiatry and a visiting professor at the School of Zoology at Dar es Salaam University, Tanzania. She is the author of *In the Shadow of Man.*

JOHN NAPIER is Visiting Professor of Primate Biology at Birkbeck College, University of London. A frequent broadcaster, he is also the author of several books, including *A Handbook of Living Primates* and *The Roots of Mankind.*

THE QUEST FOR MAN

Edited by Vanne Goodall

With contributions by Barbara Bender

Theodosius Dobzhansky

Irenäus Eibl-Eibesfeldt

David A. Hamburg

Jane van Lawick-Goodall

John Napier

PRAEGER PUBLISHERS
New York

Car enfin qu'est-ce que l'homme dans la nature? Un néant à l'égard de l'infini, un tout à l'égard du néant, un milieu entre rien et tout. Infiniment éloigné de comprendre les extrêmes . . . également incapable de voir le néant d'où il est tiré, et l'infini où il est englouti.

(Pascal, *Pensées*)

For, after all, what is man in nature? A nothing compared with infinity, all when compared with nothing. A middle point between nothing and all. Infinitely far from understanding the extremes. . . . equally incapable of seeing the nothing whence he emerges and the infinity in which he is engulfed. (*Trans.* J.-C. Peissel)

Published in the United States of America in 1975
by Praeger Publishers, Inc.
111 Fourth Avenue, New York, N.Y. 10003

© 1975 by Phaidon Press Limited

Library of Congress Cataloging in Publication Data
Goodall, Vanne Morris.
  The quest for man.
  Bibliography:
  Includes index.
    1. Man. 2. Anthropology. I. Lawick-Goodall, Jane,
Barones van. II. Title.
GN24.G63   301.41′1   72-79550
ISBN 0-275-49770-4

Printed in the United States of America

Contents

To the memory of L. S. B. Leakey

The idea for this book was inspired by the outstanding work of my late friend and colleague, Louis Leakey, to whom it is most fittingly dedicated.

For nearly half a century, Louis devoted his boundless energies to the task of seeking new fossil evidence to extend man's knowledge of his prehistoric past, in the belief that a more comprehensive understanding of the evolutionary processes at work in our planet will help man to safeguard his species from extinction and direct its course towards a better future.

The many contributions Louis made to palaeontology, archaeology and anthropology, and the far-sighted vision which gave his work meaningful purpose, are a source of inspiration to many who are now carrying on his research and have won for him a lasting place among the illustrious pioneers of prehistory.

Julian S. Huxley, F.R.S.

Introduction

Vanne Goodall

THE study of man, his physical structure and behaviour, his hopes for an extension of life after death, and the mystery surrounding his origins have intrigued and baffled generations of philosophers, scientists and artists.

Today, the quest for a more profound understanding of the nature of man, and of the structural, cultural and functional changes through which he has passed during the course of his evolution is being undertaken by an unprecedented number of inter-related disciplines, including palaeontology, archaeology, medicine, genetics and the behavioural sciences.

The Quest for Man is an attempt to draw together some of the skeins of modern scientific and metaphysical thought concerning human evolution. Each of the six contributors is a leading expert in a particular field of enquiry, whose work is complementary to that of the others, so that the book as a whole reveals an integrated picture of mankind, from which it may be seen how man's ancient past relates to his present situation.

The editor sets the scene with a brief historical summary of man's restless quest for the source of his origins and his place in nature, emphasizing some of the major discoveries made in the Western world to which man owes his present perspectives of space, time and human development.

Professor David Hamburg, a research psychiatrist, discusses the paradox of 'ancient man in the twentieth century', and the extent to which he is biologically and socially prepared for coping with the difficulties that he has created in the world today. Some of the old ways of fostering survival are proving ineffective. New ways of meeting the needs of the present may be discovered by searching for a better understanding of the basic motivations for survival that are rooted in the soil of human origin.

From the heterogeneous world of the twentieth century, *The Quest for Man* turns back to the prehistoric times that cradled the slow evolutionary development of the primates. As an anthropologist, Professor John Napier is primarily concerned with the structural development of man. The quest for his physical origins follows the fossil trail from non-human to human primates, down the culs-de-sac which end with the extinction of the three known subspecies of man, and so to *Homo sapiens* himself. The search is by no means at an end, for primate fossils are not easy to find, and it is on slender and often fragmentary fossil evidence that prehistorians have had to draw the complex outline of the human tree of evolution.

The assumptions of the archaeologist, like those of the palaeontologist, are founded on concrete evidence. Dr. Barbara Bender reviews the uneven progress of the cultural evolution of the human species. For something like three million years, until approximately 20,000 years ago, early man made little change in his simple, stone tool kit. Then there was a rapid development of technical skills associated with the late hunting-and-food-gathering societies leading, some 11,000 years ago, to the agricultural revolution which was to determine the future of the human species.

There is concrete evidence, then, of man's structural and cultural evolution, and until recently it formed the basis for scientific interpretations of early man's way of life. Now, attempts are being made to reach an understanding of his behaviour by the study of his closest living relative—the chimpanzee. In her contribution to the book, Dr. Jane van Lawick-Goodall discusses the long-term study of wild chimpanzees in the Gombe Stream National Park. The possibility of parallel evolution in chimpanzee and man is not ruled out, but the striking similarities observed between the two species suggest that they once diverged from a common ancestor. This being so, it is probable that those characteristics shared by modern chimpanzee and modern man may have been present in that common ancestor and therefore in early man. Studies of chimpanzees may help to highlight traits which are unique to man, and suggest how the precursors of such patterns of behaviour might have been shaped by evolutionary and social pressures towards their uniquely human form.

Ethologists are also seizing swiftly vanishing opportunities to throw fresh light on the behavioural patterns of early man by scientific research on some of the few hunting-and-food-gathering societies which have survived into the twentieth century. Professor Eibl-Eibesfeldt discusses his current study of the Bushmen of Botswana, whose way of life, although adapted to modern conditions, is still basically that of a hunting-and-food-gathering people, comparable to that of their remote ancestors. He suggests that only such a long period of time as that during which early man lived at the economic level of a hunter could have accommodated the genetic shaping of his behaviour. Scientific studies of people living in similar conditions may yield information as to what extent man's nature was predetermined by his genetic heritage and how far it may have been influenced by cultural environment.

Man may be curious as to why his species looks, moves and behaves as it does, but his primary concern is with himself as an individual, with his quest for a meaning and a purpose in his life and his hopes for some continuity of it after death. In a broadly based review of human evolution, Professor Theodosius Dobzhansky, a geneticist, discusses some of the many controversial theories which have been advanced through the ages in the course of man's quest for an acceptable self-image. He suggests that man is still evolving—an unfinished experiment—and that within the limits imposed by the laws of nature, he is called upon to mould his future in accordance with his ideas of goodness and beauty.

As a postscript to this introduction, I would first of all like to acknowledge a deep debt of gratitude to the late Dr. Louis S. B. Leakey for his help and co-operation in the original planning of *The Quest for Man*. The book has been dedicated to him, with a tribute by Sir Julian Huxley, to whom I would like to express my most sincere and grateful thanks. Next I would like to express my thanks to the authors, whose contributions have made the creation of this book possible. I am especially grateful to Professor David Hamburg for the helpful suggestions he made when going through my typescript and for his unwavering support and help during the preparation of the book.

I cannot adequately express my gratitude to my daughter, Jane, for without her unstinting help and encouragement it would have been infinitely harder to have carried the project through. I would also like to express my thanks to Mr. Timothy Hooker, who gave up valuable time to assist me in going through all the typescripts, and for the advice he tendered.

I am greatly indebted to Jean-Claude Peissel. The idea for the book was his and it was he who first approached and enlisted the help and support of Dr. Leakey. Louis's time schedule did not allow him to do more than give the project his moral support and he asked me if I would undertake its general editorship. Throughout the difficult days of structuring the book and trying to meet impossible deadlines, I have at all times been grateful to Jean-Claude for his forbearance, patience and close co-operation in each stage of the venture. I would also like to thank Rosemary Saunders for her unfailing good humour in dealing with seven far-flung contributors and for her hard work in moulding their work into a readable whole.

This book has been designed as an entity and is not merely a collection of essays. I would like to emphasize, however, that each author is personally responsible only for the opinions and interpretations offered in his or her own work, and cannot be held accountable for the views expressed by other contributors.

Although in some cases the choice of illustrations was made by the authors, on behalf of all the contributors and myself I should like to thank Mrs. Nomi Rowe for the research and collection of most of this material. For technical reasons the captions, although checked by the authors, are entirely the responsibility of the publishers.

In addition to those individuals and institutions acknowledged at the end of the book, the authors and publishers would especially like to thank the following for their help: Lady Collins, the Commonwealth Fund, John Crocker, Dr. Roger S. Fouts, Grant Heidrich, Hugo van Lawick, Caroline Tutin, and, in particular, Leighton Wilkie, who has given invaluable assistance and support to this project.

Ill. 1. Michelangelo's superb *Creation of Adam* captures the traditional biblical concept of 'special creation'. Like many other creative geniuses of the Renaissance, whose work was inspired by a deep religious experience, Michelangelo's painting reflects a classical appreciation of physical perfection and a visionary quality which was far in advance of his times.

1 Setting the Scene

Vanne Goodall

Ill. 7 It is now many centuries since one of the earliest and best-known attempts was made to account for the diverse forms of life on Earth. In Genesis, the first Old Testament book of the Bible, it is written of mankind: 'Let them have dominion over the fish of the sea, and over the fowl of the air, and over every living thing that moveth upon the earth.' This ancient command has been fulfilled. Man is now the most dominant form of life on the planet Earth, which has cradled the slow processes of his evolution. In acquiring his present status, he has used parts of the intricate mechanism of his brain and the delicate manipulative ability of his hands to assess and utilize the resources of his globe.

Ills. 2–5 Alone of those with whom he shares his earthly heritage, man has long since developed the means to communicate abstract thought through language, art, music and the written word. He also possesses the unique distinction of being fully conscious of his own existence and is aware that the unpredictable span of his life must inevitably terminate in death.

Ill. 24 Man's concept of time embraces past, present and future but, although it has been greatly expanded by scientific discoveries relating to the geological age of the Earth, it is still confined by the boundaries imposed by a finite mind. It is currently estimated that there has been life on this planet for approximately 3,300 million years. The very recent appearance of *Homo sapiens* in recognizable modern form—a mere 40–50,000 years ago—suggests that man is still in the early stages of his development and that, providing the processes of evolution are infinite, he may one day acquire the ability to perceive the universe in its true perspective.

Our early ancestors left no written records to help the prehistorian in his interpretation of their way of life. He must do the best he can with the evidence provided by the fossil material, the stone tools, and the paintings on the walls of caves. The historian consults the written records of man in his quest for knowledge—a quest made possible by the invention of phonetic writing in *c.* 3100 BC. The Sumerians were probably the first people to *Ill. 4* advance a system of communication by symbol, ideogram and picture story into a fully-developed form of writing, by which it was possible for a written sign to represent a sound. In time, the principles of phonetization were adopted by other great urbanized civilizations of the ancient world with momentous and far-reaching consequences to the whole field of human progress. The records of their scribes, priests, professional letter-writers, and merchantmen form the nucleus of the history of mankind.

Basically, the behaviour of the early urban societies was similar to that of their descendants of today. Comparatively safe at last from the terrors and dangers besetting the life of a nomad, the city-dweller and the agriculturist were able to reap the material benefits derived from the technical skills which had been developed by men of the New Stone Age—skills which were to determine the whole future of the human species. Homes were built of stone or brick, families were brought up in the tradition of their ancestors or of the community in which they lived. The same motivations which are extant in the twentieth century AD were present then. The urge existed to explore the unknown territories which surrounded their small world, to trade with the people they discovered on their journeys, and to satisfy their need for self-expression in creative works of singular beauty.

And throughout this vitally formative age, it is impossible to overestimate the power of the religious impulses which shaped and inspired man's day-to-day activities. Mythological legends and rituals passed by word of mouth from former times, *mana*, animism and ancestor worship were woven into the beliefs he cherished. He believed that the swing of the sun and the moon, the rumble of approaching thunder, the giant clap overhead, the race of the floods which threatened his safety or irrigated his lands, were due to the activities of a galaxy of good and evil spirits, gods and goddesses, demons and monsters. Even man's earthly rulers were looked upon as gods, or their ambassadors, and their laws were obeyed without question.

Inevitably, perhaps, these early pages of history record not only the progress of man's religious, artistic and technical development, but also the aggressive violence of which the human species is capable. The descendants of the men who had competed with the predators of forest and savannah for food and shelter now began to wage highly organized territorial warfare

Ill. 6

Ills. 2–5. The biblical narratives of the Creation and the Flood are similar to those found in earlier Mesopotamian sources. These were passed by word of mouth from generation to generation and were probably set to music. The golden bull's head (opposite, far left) decorated a lyre not unlike the one shown on the Royal Standard of Ur (opposite, left). One of the earliest texts containing the stories of the Creation and the Flood was found by Sir Austin Henry Layard at Nineveh in 1850. It was deciphered by George Smith, who found it to be part of the epic of Gilgamesh, the legendary hero of prehistoric Mesopotamia—two-thirds god and one-third man (right). The fragment of the tablet (above) tells the story of the Flood.

upon members of their own species. With monotonous regularity and ever-increasing savagery, the pattern then formed by man of peace and war, of construction and destruction, of progress retarded by prejudice, has persisted until the present day.

The Iron Age, infiltrating into the Bronze Age, witnessed the disintegration and destruction of all the great civilizations of the ancient world except that of China, and the shift of the centre of human civilization from the Middle East to Greece and Rome.

During the last millennium before Christ, man reached a new stage in his development. He began using his ability to make objective observations. Blind, unquestioning acceptance of the old evaluations of life forms, natural phenomena and the source from which all had emanated, were challenged by reason. Where literate cultures had been established in countries as far apart as Asia, China, Persia and Greece, revolutionary ideas generated, spread and were diversified. Nowhere did this new freedom of thought find a richer breeding ground than in the sophisticated atmosphere of Greece. Encouraged by such pioneering ideas as those of Thales of Miletus in the sixth century BC, its sudden, rapid expansion remains one of the most outstanding phenomena in human development. Under the tutelage of such intellectual giants as Anaximander, Pythagoras, Socrates, Plato, Aristotle and other great philosopher–scientists, the Greeks began to seek new and rational explanations for the meaning of life.

In the third century BC three nation states—Syria, Macedonia, and Egypt—were established under Greek rule, and the Ionian culture fused with that of the Near East and became known as the Hellenistic civilization. New fields of science were now continually being opened up. Hippocrates had already fathered the birth of modern medicine in the fourth century BC, and a reasonable atomic theory had been advanced by Leucippus and Democritus. The last three hundred years of this millennium witnessed an astounding number of fresh discoveries. Without any technical aids, Eratosthenes measured the circumference of the world with only marginal error. Specific gravity was discovered by Archimedes, and Aristarchus advanced the theory that the earth and other planets revolve round the sun. Herodotus, the Greek historian, while crossing the North African deserts, recognized the significance of the fossil when he concluded that the fossilized shells he saw were the remains of creatures left there in the long distant past when the sea had retreated from the area. Anaximander, too, anticipated modern knowledge by advancing the theory that man had evolved from fishes. Mathematics, physics and astronomy all have their roots in this great age of learning. In the world of art and architecture, poetry and drama, the Hellenistic culture set a high standard of majestic grandeur. No civilized peoples have ever made a greater contribution to the free expansion of the human mind or to the ideals of beauty and tolerance.

This evidence of the potential of the human brain runs parallel in history with the list of the wars, which then, as now, reveal the essential immaturity of the basic human behaviour pattern. In spite of having clear foreknowledge of the consequences of organized warfare—the loss of food supplies, the famine and pestilence, the misery, devastation and want, the mutilation of bodies—the dominant members of the Greek and then the Roman societies plunged their fellow human beings again and again into the irresponsible

Ill. 6. From the earliest times, man has believed in the power of magical and religious forces from which he has sought favours through acts of worship and conciliation. It is probable that the Magdalenian hunters of Upper Palaeolithic times were influenced by such beliefs when they painted the deep recesses of the caves of the Dordogne, France, with figures of the animals they hoped to capture or whose fertility they wished to encourage (see *Ills. 38, 104*). The powerful painting in the Shaft of the Dead Man at Lascaux of a bison, eviscerated by what seem to be spears or arrows, also shows an ambiguous figure with a bird's head (opposite). This has been interpreted by various authorities as probably representing a Shaman and as having some totemic and magical significance. It is exceptional in that human beings are rarely represented in cave art.

14

savageries of war. It is perhaps one of the contributing factors to the non-extinction of the human species that his creative and religious impulses survive the evils of prolonged human conflict.

When the Roman Empire rose to dominance, the Hellenistic culture was absorbed. Fortunately for posterity, the conquerors were eager to be tutored by the conquered, so that wherever the Roman Legions bludgeoned their way across the Western world, they took with them a little of the light which had been shed by the creative genius of the Greeks.

It was a Roman poet, Titus Lucretius Carus, inspired by Epicurus, the Greek philosopher, who was one of the first to record a recognition of the struggle for existence which plays such an important part in natural selection.

Ill. 9 His famous masterpiece, *De Rerum Natura*, was published *c.* 53 BC and in his discourse on the emergence of life, he wrote: 'Every species that you now see drawing the breath of life has been protected and preserved from the beginning of the world either by cunning or by prowess or by speed. . . . But those that were gifted with none of these natural assets, unable either to live on their own resources or to make any contribution to human welfare, in return for which we might let their race feed in safety under our guardianship—all these, trapped in the toils of their own destiny, were fair game and an easy prey for others, till nature brought their race to extinction.' More than two thousand years have elapsed since the death of Lucretius, but it is still possible to reconcile his teachings with the findings of modern science.

It was into this historic period of rapid, intellectual development paralleled by bitter warfare that Jesus of Nazareth was born. The Roman Empire was then at its zenith and covered an area of some three million square miles, from Britain to the Sudan, from the Atlantic to the Arabian desert; Roman traders fanned out to Denmark in the north, and to India, China and even Malaya.

Jesus, we are told, was the son of a carpenter, who grew up in a little town in Palestine. Practically nothing is known about him until he reached the age of thirty, when he chose twelve men to assist him in his work, which was dedicated to the teaching of a religion which sought to promote love and peace among the peoples of all nations. There is something awe-inspiring in the reflection that, although Jesus left no written words behind him, the records of his spoken words, which could all be bound within the covers of a single slender volume, have been translated over the centuries into many languages, played an important part in shaping human history, and given inspiration and meaning to the lives of billions of human beings. In seeking to give expression to the power of religious inspiration, whether it has been found in the words of Jesus, or of Confucius, of Siddhartha the Buddha, or of Mohammed, or under the influence of any other faith by which men live, human beings have created some of the world's most moving music, art, and literature, and performed acts of great physical and moral courage.

It is believed that Jesus was only in his early thirties when he was betrayed by a 'friend' and crucified by the Romans. Until AD 313, when an Edict of Toleration was granted to his ever-growing number of followers, they periodically suffered torture and death. Eventually, in AD 389, Christianity was adopted as the official religion of Rome.

In the Far East the Chinese Empire was now entering a period of temporary decline, and in the West the power of the once mighty Roman world was being insidiously undermined by a corrupt and luxury-loving ruling class and the demands made on the economy by a starving proletariat. In this decadent and materialistic society there was little strength left to combat the evils of inflation and the ravages of barbarian invaders. Decimated by the plagues which swept across Europe, threatened from within and without, the Roman Empire was fast approaching the total disintegration which marked the cultural deterioration of the Dark Ages.

And while the whole structure of the Western world was undergoing the convulsive upheavals of political and social change, the Church* was extending its influence to the boundaries of the old Roman Empire and far beyond. The religion taught by the early Christian missionaries—the all embracing religion of love and forgiveness which Jesus had given to the world—was new and exciting. It offered comfort and hope for poor and rich alike and for the intellectual there was the promise of discovery—the end of the quest for an answer to the fundamental questions men ask themselves about the human species.

Ill. 7. Opposite: this fourteenth-century miniature shows God presenting Eve to Adam. It is a charming illustration of the medieval interpretation of the biblical concept of man's place in nature—having 'dominion over the fish of the sea, and over the fowl of the air, and over every living thing that moveth upon the earth'.

* The Church, used in the context of this book, refers to the Christian Church as a whole, irrespective of sect.

Le premier chapitre du premier livre p̄le de dieu le souuerain

En commençant a declairer aucunes choses des proprietez et des natures des choses tant espirituelles comme corporelles nous prendrons nostre commencement a cellui qui est commencement z fin de tous biens Et au commencement nous requerons la de du pere de lumiere de qui vient tout bien et tout don qui est parfait Sicque cellui qui enlumine tout homme qui vient en ce monde et qui de tenebres reuelle les choses parfondes et les choses

nuidees amaine a lumiere vueille mener a bonne consummacion ceste petite ocuure que a sa loenge et au proufit de ceulx qui la liront soy recueillie et non pas sans labour de diuers dis des sains et des prophetes Le ij chapitre parle de lumere de la diuine essence et de la pluralite des persones.

Il est donc personnes Si comme dit innocent vn seul vray dieu pardurable sanz mesure muable tout puissant le pere le fils et le saint esperit trois persones en vne essence vne substance et vne nature simple en toutes manieres Le pere nest de nullui Le fils est du pere tout seul Le saint esperit est du pere z du fils sanz commencement z sanz fin Le pere est entendement Le fils est plaisant

Ills. 8, 9. Plato taught in Athens around 385 BC in a garden dedicated to the hero Academus (opposite). His 'academy' influenced the philosophical thought, not only of his time, but also of the Renaissance, when his search for a new and rational explanation of the meaning of life was discussed and pursued. Like the ideas of Plato, those of Lucretius on the origins of man were re-investigated during the Renaissance and provided one of the foundations upon which the 'new' sciences were developed after the Middle Ages. This edition of Lucretius' *De Rerum Natura* (below) was published in 1565 in Antwerp by the great scholar and printer, Christopher Plantin.

In the first century AD, history records that Paulinus, one of Augustine's followers, was discussing the merits of the Christian faith with the wise men of Northumbria when one of them burst forth: 'So seems the life of man, O King, as a sparrow's flight through the hall when you are sitting at meat in wintertide, with the warm fire lighted on the hearth but the icy rainstorm without. The sparrow flies in at one door and tarries for a moment in the light and heat of the hearth fire and then, flying forth from the other, vanishes into the wintry darkness whence it came. So tarries for a moment the life of man in our sight, but what is before it, what after it, we know not. If this teaching tells us aught certainly of these, let us follow it.'

The followers of the Christian faith, inspired by the exemplary lives of the monks, soon multiplied and the monasteries and cathedral schools established by the Church became the centres of education. Many cultures owe a debt to the early Christian monks, for it was due to their industry that the torch of learning was kept alive throughout the Dark Ages of Western history, and that much of the classical literature which might otherwise have been lost to posterity was carefully preserved.

Paradoxically, while the teachers of Christianity preached a religion founded on a belief in the redeeming qualities of love and peace, mercy and forgiveness, they were empowered by the authority of the Church to enforce its dogma, when necessary, by means of torture and death. Heresies invoking intellectual controversy were dealt with more tolerantly: the works of the man, instead of his body, were consigned to the flames. Any views supporting those expressed by Lucretius that human beings had once 'lived out their lives in the fashion of wild beasts roaming at large', hunting 'the woodland beasts by hurling stones and wielding ponderous clubs', were regarded as a direct challenge to the doctrine of 'special creation' and strictly suppressed. This doctrine, based on the Church's interpretation of the Old Testament version of the Creation, maintained that all species, including man, had remained exactly as God had created them in the beginning. This belief in the immutability of species restricted every scientific approach to the study of man's origins in Christian communities all over the world for a great many decades.

The temporal and spiritual influence of the Church now dominated the Western scene and the growth of intellectual thought was retarded by the strictures of prejudice and dogma, fear and superstition, until it was set free again in the Italian Renaissance, which followed the Fall of Byzantium in 1453. During this re-birth of the human mind, ideas relating to the arts and the sciences, which had been confined during the Dark Ages, were re-investigated and developed. Christianity was interpreted by a galaxy of brilliant painters and sculptors, including the immortal Michelangelo and Leonardo da Vinci. Greek and Latin literature, which had been treasured in the monasteries, was brought to light and a zest for classical education spread rapidly from Italy across the continent to England. It was a time of tremendous creative activity, a transition period of major importance to the human species, when the medieval outlook on life underwent some of the radical changes which have led to our present perspectives of space, time and human development.

In the last three hundred years man has been subjected to a series of major psychological shocks. The first of these to concern this summary occurred

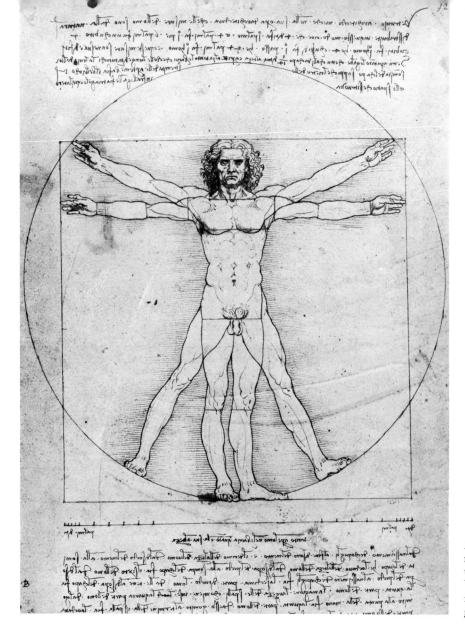

Ill. 10. With the renewal of learning man became the measure of the universe, and this illustration of Vitruvius' proportions of the human figure by Leonardo da Vinci seems to give graphic shape to this philosophy.

Ill. 186

Ills. 11, 12

during the end of the sixteenth and the beginning of the seventeenth centuries, when the work of the great astronomers—Galileo, Copernicus, Kepler and Newton—destroyed the medieval belief that the planet Earth was the centre of a fixed and perfect universe, ordained and directed by a divine Creator. In place of this Ptolemaic concept, man now began the slow process of adjusting his mind to the new vision of a universe that was vast beyond the powers of his imagination, with his own world, no greater than a speck of dust, revolving in space around the sun.

While man struggled to assimilate this new perspective of space, Archbishop James Ussher, basing his calculations on biblical chronology, announced in 1593 that the world had been created in 4004 BC. Into this rigid framework of time the Archbishop had fitted the Old Testament story of the Creation, the Noachim Flood and the subsequent history of the human species.

18

This erroneous assumption that the world was barely six thousand years old became an accepted tenet of the Christian faith, hampering and confusing all progressive ideas concerning man's origins.

It is instructive to reflect that, but for the use of the word 'day' in the much translated version of the story of the Creation in the first two chapters of Genesis, the many controversies that arose owing to the limitations imposed by Ussher's time-scale would probably never have taken place. The late Dr. Louis Leakey, one of the greatest prehistorians of all time, considered that 'Genesis was written in the language of poetry and not in the language of science. Having accepted this point, we can look at Genesis afresh. When we do so, we find that the writers of it had the main sequence of events about the world remarkably correct, but we must treat the six days of Creation as representing not *days* in our present sense of the word, but as *periods* of time.'

Unfortunately, the Archbishop only allowed a literal six *days* for the Creation of the world, and it was more than a hundred years before his calculations were seriously challenged. In the meantime, although several notable attempts were made to establish the vital role of the stone tool and the fossil as a means of interpreting a long distant past, all had met with failure. For centuries, stone tools found in many parts of the globe had been regarded as curiosities of nature and in particular as 'thunderbolts'. Theories which were put forward from time to time, that these little shaped stones were man-made and could bear witness to the cultural activities of some primitive form of man who was ancestral to Adam, were condemned and suppressed by the Church.

Efforts to establish a scientific interpretation of fossils shared the same fate. It was now nearly two thousand years since Herodotus had recognized the organic origin of the fossil but, in spite of the periodic support his view had been given by such men of genius as Leonardo da Vinci, it had been denied official recognition because of the restriction of Archbishop Ussher's time-scale.

Ills. 11, 12. In the realms of geography, cosmology and other sciences, as well as in the arts, the savants of the Renaissance investigated, criticized and developed many of the theories they found in the classical literature which had survived the Dark Ages. Ptolemy's earth-centred universe (below) was gradually rejected in favour of the theories of Galileo, Copernicus (see *Ill. 186*) and later Kepler and Newton. It was Ptolemy's knowledge of geography, however, as shown in the fifteenth-century Italian re-drawing (below right), which inspired travellers, such as Marco Polo, to explore the world for themselves.

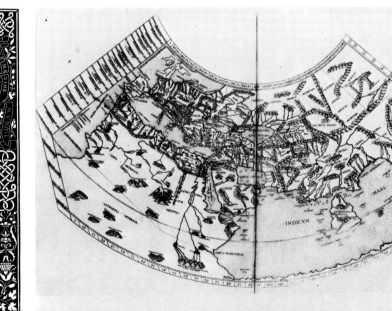

But the ever increasing accumulation of information concerning the geological story of the earth persuaded the enlightened thinkers of the late eighteenth century that its real significance could no longer be ignored. In order to find some explanation which would be acceptable to clergy and scientist alike, the Old Testament was consulted, and the Diluvial theory put forward. This claimed that fossils were the remains of creatures which had been drowned in the Noachim Flood and was strongly supported on all sides until new data proved it to be untenable. Geologists had now discovered groups of fossilized animals, each differing from the others, in successive geological strata, indicating that they had lived at different geological periods of time and could not have been drowned in one and the same universal deluge. Reports too, from such far-flung areas as Asia, America, Africa and Europe, were claiming that fossil forms ancestral to the

Ill. 13. As man's awareness of himself and of his environment developed, so did his preoccupation with Time. This subtle composition captures the elusive quality of man's quest for a better understanding of his past, present and future.

fauna of each had been found in geological deposits laid down before the biblical Flood could have taken place. One way or another, it seemed as though the ticklish problem of how many species Noah had been obliged to pack into his Ark, which had been the cause of so much speculation in the past, could now be resolved, since only a miracle could have enabled him to collect individuals of species from all over the world and redistribute them after the Flood.

Ill. 16

The French palaeontologist, Baron Georges Cuvier (1769–1832), in a monumental effort to fit this new information into biblical chronology, advanced his Catastrophic theory. This suggested that there had been a series of catastrophies, each followed by a period of calm during which God had restocked the world with life. But even Cuvier found it impossible to fit three additional creations into Ussher's time-scale and, following the lead given by his eminent colleague, Comte Georges de Buffon, he pushed the date for the Creation of the world backwards in time, adding some 80,000 years to the age of the Earth.

Man has never found it easy to adjust his mind to the reception of new ideas. Fear, superstition and prejudice more often than not will urge him to cling to the traditional beliefs of his family, his community and his culture. These long drawn-out controversies concerning man's place in nature, which arose as a result of an obdurate adherence to Archbishop Ussher's time-scale, clearly reveal how prejudice, smothering the dictates of reason, retarded the progress of man's quest for the true source of his origins, just as it has done in so many other fields of human endeavour throughout the ages.

With the close of the eighteenth century, the scientific approach to the study of man was already beginning the slow process of emancipation which was to lead to the freedom it enjoys today. The opening decades of the nineteenth century witnessed a widespread enthusiasm for fossil-hunting, and a more logical assessment of the evidence which was brought to light. But although fossilized remains of extinct animals were given official recognition, all claims to have discovered the remains of man, in deposits laid down in Europe before the Noachim Flood, were still greeted with ridicule and hotly criticized. There can be a no more striking demonstration

Ills. 14, 15. In the eighteenth century, new scientific information was beginning to challenge the Christian belief in a 'special creation'. Recognition of flint implements claimed to be of human origin and found in geological deposits laid down before the Flood (such as the Lower Palaeolithic hand-axe (above) found by John Frere at Hoxne in Suffolk) meant a rejection of Archbishop Ussher's calculation of the world's antiquity. Some scientists, however, still strived to fit the discoveries of fossil bones and man-made implements into biblical chronology, with the result that Scheuchzer saw in this fossilized giant salamander that he discovered in 1726 the 'afflicted skeleton of an ancient sinner' drowned in the Flood (right).

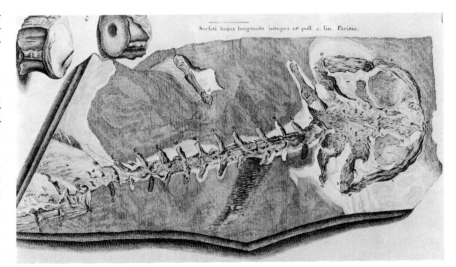

of the strength of the religious prejudices which clouded the judgement of so many of the early nineteenth-century scientists of the Western world than their rejection of a prehistoric ancestry for man, while at the same time accepting the existence of a prehistoric fauna. Only a minority of the investigators into man's past, undaunted by the heat of the controversies aroused by their work, persistently maintained their firm conviction that the history of man stretched back beyond the boundaries that had been laid down for it by the Church. Listed among the names of the indomitable pioneers of the science of prehistory are those of John Frere and Boucher de Perthes, Father J. McEnery, Dr. Schmerling, and Edouard Lartet.

Geological evidence for an even greater expansion of the age of the Earth was now overwhelming and in 1830 Sir Charles Lyell, an English geologist, (famous for his synthesis of the doctrine of uniformitarianism, which maintains that the forces of nature have always worked in a uniform manner) published *The Principles of Geology*. In this book, which became a landmark in the annals of geology, Lyell measured the age of the Earth in millions of years, instead of the thousands allowed for by his predecessors.

Once more, as the impact of Lyell's work began to be felt outside scientific circles, man began the slow process of adjusting his sense of perspective, this time not to the inconceivable and frightening vista of eternal space, but to the awe-inspiring concept of the antiquity of his world.

Only such a vast span of time as Lyell allowed for the age of the Earth could have accommodated the slow and gradual processes of Darwin's theory of evolution. The conception of evolving life was not new. Revolutionary ideas which were in direct opposition to the immutability of species had been advanced by the great thinkers of the eighteenth century, but it was Charles Darwin who finally gave man a new and logical theory which would account for the long succession of divers forms of life on earth. *The Origin of Species by Means of Natural Selection of the Preservation of Favoured Races in the Struggle for Life* was published in 1859 and was based on Darwin's belief that all species are mutable and can trace their ancestry back to the lowliest forms of life. This theory was seen as a direct threat to the doctrine of special creation and Darwin was attacked with fanatical fury. The idea that a wise and loving God had fashioned man in his own image and therefore in a permanently immutable form was still a cherished and important tenet of the Christian faith, and many devout scientists, while recognizing the genius of a theory which could withstand the most acid tests of logic, still rejected it for purely religious reasons.

Twelve years later, Darwin published *The Descent of Man and Selection in Relation to Sex*. 'Although man has risen to the summit of the organic scale and developed a godlike intellect', he wrote, 'he still bears in his bodily frame the indelible stamp of his lowly origin.'

Wherever the impact of these revelations was felt by the general public it was followed by a profound sense of shock. For centuries, man had been flattered and uplifted by the assumption that he was a special creation. The idea of having developed from some lowly, pre-existing form was not only revolting but horrifying, because of the unpleasant suspicion that perhaps Darwin was right.

By the beginning of the twentieth century there was a growing tendency for the Church to reconcile its teachings with the new concept of human

Ills. 16, 17. It is thanks to the pioneer work of such men of genius and courage as Georges Cuvier (above) and Charles Lyell (below) that twentieth-century man was able to develop an ever-increasing awareness of his own antiquity and that of his planet.

evolution. Prehistory, no longer regarded as a hobby for cranks and amateurs, had become a recognized branch of science. Inspired by the immense possibilities opened up by Darwin's theory of evolution, the great pioneers of prehistoric field work, Eugene Dubois in Asia, Raymond Dart in South Africa, and later, Louis Leakey in East Africa, were beginning the systematic investigation of carefully selected sites which was to revolutionize man's knowledge of his origins.

Meanwhile, phenomenal advances were being made in other branches of science and the rapid multiplication of technical skills was shaping the future of twentieth-century man. Global increases in industrialization, speed of communication, and the availability of knowledge to a greater number of people were leading to the merging of cultures and the growing tendency to negate the individual. The prophecies of science fiction were beginning to come true: man was preparing to launch himself into outer space.

Once again the history of the first half of the twentieth century records a familiar pattern of human behaviour. Not only do we read of the unprecedented rate at which knowledge of the physical world was being acquired, but also of the global holocaust caused by two of the most barbaric wars that human beings have ever waged against each other. Greed, covetousness and mass indoctrination swept nation after nation into hideous conflict. Millions of human beings, forced into the defensive against aggression, were slaughtered or maimed, economies were destroyed and lands laid bare.

The Second World War was brought to an abrupt halt in 1945 by the explosion of a hydrogen bomb over the town of Hiroshima in Japan. This was the first serious demonstration of the power of the most destructive force man has ever controlled. Shocked and sobered by the news of the devastation inflicted by a single atom bomb, the world faced the terrible knowledge that mankind was in possession of the means to exterminate its own and probably all other species of life on earth.

Ills. 18–20. In spite of a growing acceptance of his theory of evolution, Charles Darwin (portrayed above twenty years after the publication of *The Origin of the Species*) was mercilessly lampooned in the press and his theories were fair game, as these two cartoons show (right). The extreme fundamentalists' reactions to 'Darwinism' gained world-wide publicity in the famous Tennessee evolution or Scopes Trial, which took place at Dayton in July 1925.

MONKEYANA.

AM I A MAN AND A BROTHER?

THE LION OF THE SEASON.

Ill. 21. Since the end of the Second World War, man has graduated from the steam age to the space age. The rapid advances which have been made in the fields of science and technology have now made it possible for him to probe the unknown (opposite).

It was not long before the world press began to administer what was perhaps the most chastening shock man has ever experienced. It came suddenly. Pollution and the population explosion became household words almost overnight, and man was made aware that his future was not only threatened by the possibility of nuclear warfare but was also seriously jeopardized as a result of his own careless and arrogant exploitation of the resources of nature.

These two psychological shocks had a salutary and maturing influence on the mind of man. Although he may no longer bow down before the swollen race of the flood waters as his ancestors did before him, he has learnt with a new sense of humility that he must now work in harmony with and not against the forces of nature.

Ill. 22. Man's inhumanity to his fellow seems to increase in proportion to his technical progress, and even religious buildings are not immune from attack. Here, a cross, the symbol of Christianity, has managed to survive amidst the ruins of a Roman Catholic church in Quang Tri, Vietnam.

Ill. 23. Contrasting bridges span the centuries. The man on a donkey crossing a stream on a medieval bridge beneath the pylons which support a modern auto-strada in Sicily typifies how a drastic technological innovation may affect a tradi-tional way of life, creating profound tension between this and new opportunities.

26

2 Ancient Man in the Twentieth Century

David A. Hamburg

* Throughout the text, numbers refer to works listed in the Bibliography on p. 223 under the relevant chapter.

Ill. 24

HUMAN EVOLUTION AND RECENT CHANGES

THIS book reflects the extraordinary upsurge in research on the evolution of human behaviour in recent years.[9.1]* New information and ideas have emerged from discoveries involving population and molecular genetics, the fossil record, prehistoric archaeology, chemical dating methods, existing hunting-and-gathering societies, and non-human primates, especially Old World monkeys and the great apes.[39,44] Despite the antiquity of the subject matter, there has only recently been scientific investigation on any substantial scale. Although generalizations must be tentative, the viewpoint is fundamental in efforts to understand the nature of man.

The time-scale of evolution highlights a dilemma in the current predicament of human biology and behaviour.[46] There have been mammals on earth for more than seventy million years. Primates appeared early among the mammals and have been present for many millions of years. A manlike form has been present for several million years. Our own species has been in existence for about 40,000 years. Agriculture has been with us perhaps somewhat less than 10,000 years; the Industrial Revolution began about 200 years ago. So, on an evolutionary time-scale, the world we live in is mainly one that we have made for ourselves very recently.

Problems that concern us so urgently today—for example, the enormous population growth, urbanization with its difficult consequences, environmental damage and resource depletion, the risks of weapons technology—are largely products of the Industrial Revolution. Some of the main features of the contemporary environment are products of the most recent phase of evolution, much of which has taken place within the memory of living adults. Natural selection shaped our ancestors in ways that suited earlier environments over millions of years.[15] We do not know how well we are suited biologically to the world we now live in, but we must try to understand the forces that moulded our species in the past.

Evolutionary adaptation refers centrally to the reproductive success of a population—i.e., passing the genes to future generations. The reproductive success of a population is positively correlated with the reproductive success of individual members of that population, which is in turn positively correlated with the absence of intense and prolonged suffering and with effective social behaviour. I shall have more to say about this shortly. When the environment is stable in its main features over long periods of time, guidelines for behaviour emerge that are, on the average, useful for the population in meeting adaptive tasks. Such guidelines for behaviour tend to

27

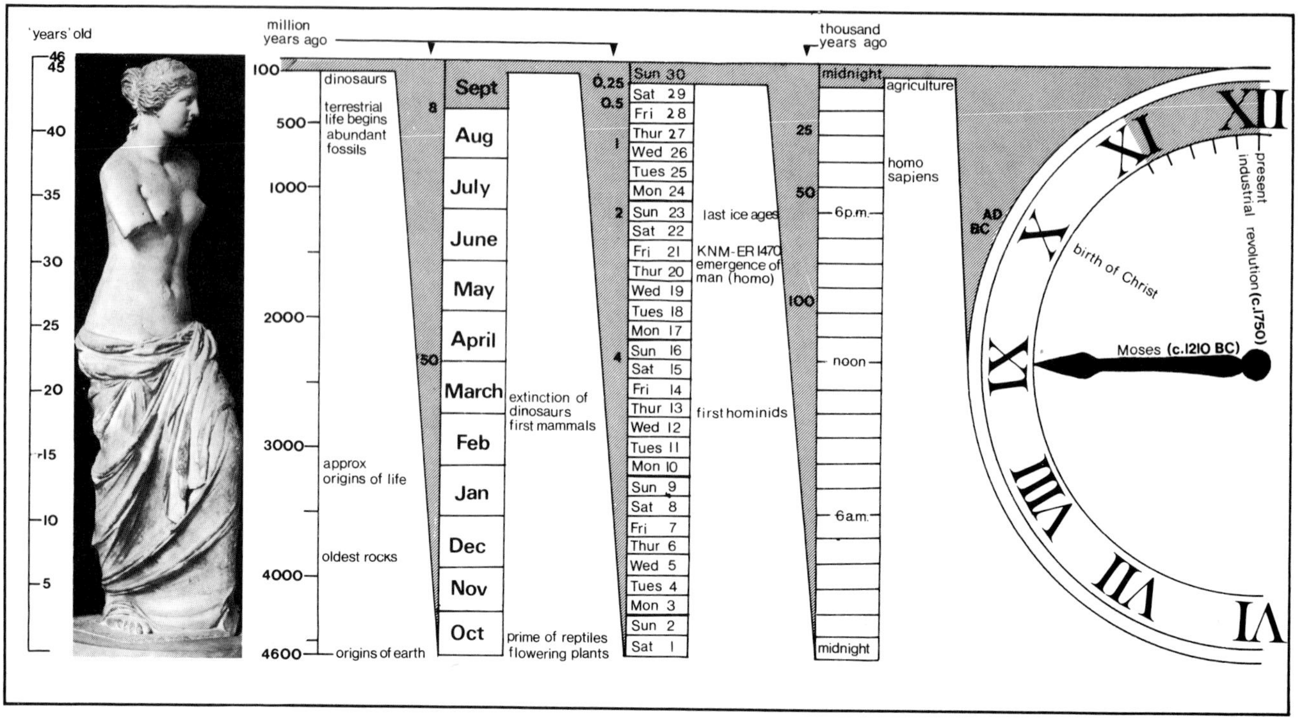

'years' old
46
45
40
35
30
25
20
15
10
5
million years ago
100
500
1000
2000
3000
4000
4600
dinosaurs
terrestrial life begins abundant fossils
approx origins of life
oldest rocks
origins of earth
8
50
Sept
Aug
July
June
May
April
March
Feb
Jan
Dec
Nov
Oct
extinction of dinosaurs first mammals
prime of reptiles flowering plants
0.25
0.5
1
2
4
Sun 30
Sat 29
Fri 28
Thur 27
Wed 26
Tues 25
Mon 24
Sun 23
Sat 22
Fri 21
Thur 20
Wed 19
Tues 18
Mon 17
Sun 16
Sat 15
Fri 14
Thur 13
Wed 12
Tues 11
Mon 10
Sun 9
Sat 8
Fri 7
Thur 6
Wed 5
Tues 4
Mon 3
Sun 2
Sat 1
last ice ages
KNM-ER 1470 emergence of man (homo)
first hominids
thousand years ago
25
50
100
midnight
6 p.m.
noon
6 a.m.
midnight
agriculture
homo sapiens
AD
BC
birth of Christ
Moses (c.1210 BC)
present
industrial revolution (c.1750)
XII
XI
X
IX
VIII
VII
VI

Ill. 24. Because the figures involved in geological time are so vast—4,600 million years—as to become meaningless, let us imagine Mother Earth as a woman of forty-six, each 'year' of her life representing 1,000 million years of time. The table shows how little is known of the first seven 'years' of her life, and the salient points in her development. In the span of forty-six 'years' man has only occupied the last ten days and *Homo sapiens* the last four hours. The beginning of our era occurred ten minutes ago and the Industrial Revolution—three real lifetimes ago—one minute ago.

be taught early in life, shaped by powerful rewards and punishments, invested with strong emotions supported by social norms. Early learning of guidelines for adaptive behaviour tends to induce lifelong commitments. On the average, over long time-spans, these strongly held guidelines prepare the young to meet the adaptive requirements of the environment by fulfilling the roles of adult life.

While biological adaptation means reproductive success of a population, the size of a population must not be considered independently of its ecological setting. Reproductive success means progeny able to utilize environmental opportunities and to avoid catastrophes. The size of the population must be related to the carrying capacity of the land, which in turn varies with technology. Much of hominid evolution has been characterized by subsistence uncertainties and threats of natural disaster. Thus, the prospect of technology that could enhance man's control of the environment has been historically attractive. In light of basic survival concerns, technology has often been pursued vigorously and has sometimes transformed society. One such transformation occurred promptly in the course of the Industrial Revolution, another between 1870 and 1900 when an acceleration in the impact of science and technology occurred; and a third acceleration has been evident since the Second World War. On the time-scale of evolution these are exceedingly recent changes, and their rate is extraordinarily rapid.

In times of such rapid technological and social change, old guidelines for behaviour tend to become uncertain or even discredited in some sectors of society—yet they persist in others. Such guidelines refer to human relationships, to man's relations with his physical environment, and his outlook on life and death. Long-established, culturally transmitted guidelines may be poorly suited to new conditions, but if they have worked well for long periods they are difficult to change. They are hard to change for an individual in his lifetime, and they are also difficult to change for the population of which he is a part. The problem of changing such behaviour patterns is substantially related to the fact that self-respect and close human relationships are dependent upon them, i.e., behaving according to established guidelines

Ill. 25 supports a sense of personal worth and a sense of belonging in valued groups. Yet the need for change in behaviour is actually greatest when drastic environmental changes are occurring. Some dimensions of culture may change much more rapidly than others—for example, a drastic technological

Ill. 23 change may disrupt the traditional way of life whose values and institutions may lag behind, clinging to a hallowed past.

For millions of years, non-human and human primate societies have been meeting the adaptive requirements of their populations. They have done this in many ways, largely depending on the environmental conditions they had to meet. Individuals have been prepared throughout a lengthy childhood development for their roles in a particular social system. Each social system supports, channels, and facilitates adaptive behaviour throughout the life cycle. In effect, each culture gives instructions on ways of meeting specific adaptive problems and sets limits of acceptable behaviour in coping with adaptive requirements. When technological changes offer hope of meeting these requirements more effectively, social systems may have to change to utilize the new opportunities. Then, there are social pressures and incentives

Ill. 26 to change child-rearing and educational practices to prepare individuals for effective roles in their modified social systems. A very rapid rate of change poses distinctive problems for the culture, since these socialization practices must alter rapidly enough to meet emerging conditions.[20]

Just as studies of animal structure and function have shown that a variety of biological solutions are possible for a given environmental problem, so,

Ill. 25. From the earliest times, man as a social animal has emphasized the membership of the group or groups to which he belonged—family, tribe, or nation. This sense of belonging to a valued group bolsters the individual's self-respect and helps him to develop as a member of the group. The ceremonies involved with coming of age have long been occasions for welcoming new members into the group and society has always accorded them great respect. Here, a young Jewish boy is being confirmed in his faith—Barmitzvah—and is seen tying the phylacteries round his arm and forehead, according to religious law.

Ill. 26. Aural learning is still an important source of knowledge of the group's heritage, thus transmitting from one generation to the next the knowledge and patterns of behaviour of the group.

too, a variety of social solutions may be utilized to meet common human predicaments. Underlying this diversity, however, are fundamental human needs. Although different cultures arrange these matters in remarkably different ways, individuals in all cultures seem to require a dependable basis for self-esteem and a sense of belonging in a valued group. In the widest variety of societies, individual adequacy in meeting adaptive tasks depends substantially on the support, guidance, and facilitation of significant other people and the society at large.

Adaptive tasks change over time, so preparation to meet them must change as well. This requires the modification of existing social institutions or even the development of new ones to assist the individual in acquiring strategies that have a reasonable effectiveness in meeting the tasks of adaptation in a world that is largely new as compared with that of his grandparents—let alone the world of early man.

HISTORICAL BACKGROUND

Boulding[2] considers the twentieth century to be the second great transition in human history. In his view, the first transition occurred 5–10,000 years ago with the advent of agriculture and the domestication of animals:

> Agriculture is a precondition of the development of civilization, because it is not until man settles down and begins to cultivate crops and domesticate livestock that he is able to develop a surplus of food from the food producer above and beyond what the food producer and his family require themselves for their own maintenance. In hunting, fishing, and pastoral societies it seems to have been hard for the food producer to produce much more than the immediate requirements of himself and his family.

He goes on to emphasize that, even when the production of a food surplus made urban culture possible, the basis of such culture was typically meagre: 'Almost all the cities of classical civilization were within a few weeks of starvation at any time, and a relatively small worsening in general conditions, in the means of transportation or in conditions of peace and war, was frequently enough to undermine the precarious foundation of civilized life.'

30

Thus, the survival-related quest for greater mastery over the environment
Ill. 27 continued. In fifteenth-century Europe, the invention of printing enormously
increased possibilities for the spread of information. In the seventeenth
century, science emerged, followed by a remarkable technological spurt in
the eighteenth century.

Brown[6] points out that a series of technological innovations in England in
the late-seventeenth and early-eighteenth centuries made possible the
development of a useful steam engine. Now, the human species had at last
a way of utilizing huge quantities of inanimate mechanical energy for a

Prelum
Afcéfianū.

Ill. 27. The invention of printing
revolutionized the spread of learning
(see *Ill. 9*). This woodcut by Dürer
is dated 1520, eighty-six years after
the invention of printing with
movable type by Johann Gutenberg.
It shows the instrument which,
more than any other, contributed to
the diffusion of the impact of the
Renaissance.

variety of adaptive tasks: 'The impact of the Industrial Revolution upon demographic history was enormous. Improvements in agriculture, transportation, and public health resulted in significant decreases in death rates. With birth rates little changed, the population of England increased rapidly. . . . During the nineteenth century in England and the greater part of Western Europe, the large variations in mortality that had been so characteristic of pre-industrial society were virtually eliminated.'

Davis[7] has called attention to crucial turning-points in the history of human population growth: 'A major shift occurred some time around 8000 BC. Up to then *Homo sapiens* had multiplied very slowly—about one-tenth of one per cent per century (assuming that the species had about 50,000 members some 400,000 years ago). The next jump in the rate did not come until around AD 1750. From 1750 to the present the average rate of growth has been approximately eighteen times that of 8000 BC to AD 1750. . . . The two big events in the expansion of the human population were the so-called 'Neolithic Revolution', which gradually added agriculture and animal husbandry to the quest for food, and the Industrial Revolution, which harnessed inanimate energy. Of the two, the second was the more abrupt and influential. The last 220 years represent less than a thousandth of human history; yet they account for more than a fifth of the total population increase. In those 220 years the earth's inhabitants multiplied four and a half times.'

Since 1750, an accelerating trend in population has developed, reaching its culmination in the period 1946–present. Davis notes that in the final tenth of the last 220 years, four-tenths of the total population increase occurred.

The recent changes in mortality, fertility, and population growth have fostered migrations on an unprecedented scale. The major flow has been from farm to city. Science-based agriculture and labour-saving farm technology greatly increased agricultural productivity and released large numbers of people from their commitment to a rural way of life.

In industrialized countries, population growth in the cities has far exceeded that in rural areas during the twentieth century. Thus, a historic reversal of population distribution has taken place very recently. This is vividly illustrated by the case of the U.S.A. At the beginning of the nineteenth century, about ninety per cent of Americans were in agriculture. By 1920, they were about evenly distributed between farms and cities. Today, less than ten per cent of the population is in agriculture.

When the Industrial Revolution began, Europe was an agrarian continent with minimal development of cities. 'In 1900 no more than a quarter of the world's population lived in cities and towns; by the year 2000 probably more than fifty per cent will be city-dwellers. No comparable period of history has witnessed such a profound transformation.'[41]

Boulding[3] points out some remarkable facts of contemporary existence: 'About twenty-five per cent of the human beings who have ever lived are now alive, and what is even more astonishing, something like ninety per cent of all the scientists who have ever lived are now alive. . . . In a very real sense the changes in the state of mankind since the date of my birth have been greater than the changes that took place in the many thousands of years before this date. . . . The great transition is not only something that takes place in science, technology, the physical machinery of society, and in the utilization of physical energy. It is also a transition in social institutions.'

Ill. 28. Opposite: this photograph shows Africa, the cradle of man, as seen from a satellite. It suggests the enormity of the distance that humanity has travelled in the several million years since ape-like men superseded men-like apes somewhere along the Rift Valley of Eastern Africa. Twentieth-century man, although he has altered his environment beyond recognition, still reacts in some ways like his earliest ancestors. Maybe it is this lack of behavioural evolution in the rapidly changing world which lies at the root of many contemporary problems.

Ill. 28

Ill. 29

Ills. 29, 30. Megalopolis is fast replacing the human scaled communities of the past (opposite). In 1900 no more than a quarter of the world's population lived in cities and towns—although the view of London by Gustave Doré already shows the social problems that overcrowding creates (right). By the year 2000 more than fifty per cent of the world's population will be city dwellers. The implications for man's behaviour of such a profound transformation are only now becoming apparent and will have to be taken into account as an ever-increasingly important factor by the social scientists and planners of the future.

ATTACHMENT

One of the major changes that has occurred since the Industrial Revolution has been the growth of enormous cities. In the industrialized countries, and more recently in the developing countries, these cities have been fostered by many currents, such as proximity to natural resources, transportation advantages, educational and occupational opportunities, and general stimulation. They are powerful magnets, especially to young men in the countryside and even to people in other countries. Often they come to the cities with only a vague image of promise regarding the way of life to be found there.

In the twentieth century, especially in the industrialized countries, an ideology has developed that individuals (or small nuclear families) should be prepared to go wherever interests and opportunities may take them—for example, leave home for educational opportunity, or move to a faraway city for occupational opportunity.

Ills. 30, 34 In the U.S.A., such moves by individuals, couples or small families are exceedingly common. Often, they involve movement to a place where the individual has no prior human relationships—no family, friends, or even acquaintances. Thus the common contemporary experience of moving to a distant urban area involves several challenging components: (1) separation from family and established friends; (2) separation from familiar surroundings; (3) contact with many strangers; (4) crowding in urban environments; (5) contact with cultural diversity—novel and strange ways of meeting life's problems.

Therefore, it is interesting to consider what is known of the evolutionary history of these dimensions of human experience. How has our evolutionary experience prepared us, biologically and socially, for the circumstances that now face us?

Ill. 31 Experiences of the sort referred to here—separation, loss, disruption of human relationships, contact with strange people, new environments, and unfamiliar ways of life—are often associated with intense emotional reactions. Evidently it is a part of the human heritage that such situations are

33

perceived as important and difficult, and thereby linked to feelings of sadness and grief, concern and anxiety, suspicion and hostility. How can we understand the evolution of such emotional response tendencies?

Emotional responses are probably so ubiquitous in the human species and so important in behaviour because they have served adaptive functions in evolution.[16,17] Most likely, they have served motivational purposes in meeting adaptive requirements, such as locating food and water, avoiding predators, achieving fertile copulation, caring for the young, training the young to cope effectively with the specific requirements of a given environment, enhancing competence in dealing with a range of environmental conditions. Natural selection has favoured those populations whose members, on the whole, were organized effectively to accomplish these tasks.

In this view, emotion reflects heightened motivation for behaviour patterns that have been crucial in species survival. Emotional responses have several components: a subjective component, an action component, and a physiologic component supportive of the action. On the whole, these are motivational patterns that have had selective advantage over a very long time-span. There is genetic variability in the substrates of these motivational patterns just as there is genetic variability in every aspect of structure and function. Natural selection has operated on this variability, preserving those motivational-emotional patterns that have been effective in meeting the requirements of survival and utilizing the opportunities encountered in the long and tortuous course of primate evolution.

Any structure or process that is adaptive on the average for populations over long time-spans has many exceptions, and may even become largely maladaptive when there are drastic changes in environmental conditions. Considering the profound changes in human environmental conditions

34

Ills. 31, 32. In today's highly mobile, urban society, the deep emotional attachments between persons, as depicted in the sketch Holbein made for the painting of Sir Thomas More and his family (right), are often jeopardized. Loneliness, even despair, may follow such interpersonal disruption (opposite).

within very recent evolutionary times, it is likely that some of the mechanisms which evolved over the millions of years of mammalian, primate, and human evolution may now be less useful than they once were. Since cultural change has moved much more rapidly than genetic change, the emotional response tendencies that have been built into us through their suitability for a long succession of past environments may be less suitable for the very different present environment.

Ill. 32

This evolutionary concept of emotion applies to the emotional experiences associated with interpersonal bonds—the feelings and actions referred to by terms such as attachment, affection, respect, and love. Such experiences were probably crucial in human evolution for the following reasons: primates are group-living forms; the primate group is a powerful adaptive mechanism; emotional processes that facilitate inter-individual bonds and hence participation in group living probably have had selective advantage for millions of years; the formation of such bonds is pleasurable for primates; they are easy to learn and hard to forget; their disruption is unpleasant and precipitates profound psychophysiologic changes that tend to restore close relations with others of the same species. Throughout the long course of his evolution, man has been a group-living form—probably characterized by intense and persistent attachments between individuals within an organized, cohesive small society. Moreover, it is very likely that the human group, throughout the history of the species, has been a powerful problem-solving instrument, facilitating the capacity of its members to cope with harsh environmental contingencies.

The remarkable work of Jane van Lawick-Goodall and her associates at the Gombe Stream Research Centre in Tanzania since 1960 has helped to clarify the role of inter-individual attachments in primate evolution.[28,29,31]

In general, her work and that of other scholars in the field of primate behaviour has diminished the gulf between human and non-human primates, especially chimpanzees. Of all the similarities, none is more interesting than the deep and enduring quality of attachments in the higher primates, especially those of chimpanzees and of humans. Gombe research has shown affectionate bonds between chimpanzee mothers and their offspring, and between siblings, that may last throughout the life cycle. Such relationships provide mutual support under stress, joint access to vital resources, and models for learning adaptive behaviour. As complex organisms have evolved, behaviour has become an increasingly important way of meeting adaptive tasks which contribute to species survival—tasks such as finding food and water, avoiding predators, achieving fertile copulation, caring for the young, and preparing the young to cope effectively with the specific requirements of a given environment. These are behavioural endeavours that contribute crucially to species survival. The role of behaviour in adaptation is not only a function of individuals, but of groups as well. This is strikingly true of higher primates. Both non-human primates and early man have been organized into small societies. These societies provide, firstly, intimate, enduring relationships with mutual assistance in difficult circumstances; and secondly, clear guidelines for individual behaviour, highly relevant to survival requirements in a given environment.

Ills. 126, 130
Ills. 132, 138

Studies of human infants have suggested pathways through which powerful and enduring attachments are formed.[47,48] Biological derivatives of our history as a species, built into the brain because they had selective advantage in evolution, favour certain patterns of interaction in infancy. In most environments, these interactions facilitate the formation of attachments. The capacities of the human infant in the first few months of life are greater than had been previously assumed.[26] The linking of a few elementary behaviour patterns early in life leads to the formation of interpersonal bonds. A visual preference for the human face, an auditory preference for the human voice, and a simple motor pattern (the smile) come together in such a way as to elicit caretaking, affectional responses from the mother. They facilitate a transaction between mother and infant that deepens the bond between them over time. This view is similar to Bowlby's formulation of attachment, centring on infant behaviour patterns that maintain proximity with the mother and elicit nurturant responses from her.[4,5] These patterns include following, clinging, crying, calling, greeting, and smiling. They function as adaptive mechanisms and have been maintained because of their survival value.

Ill. 35

Another evolutionary aspect of emotion is that individuals avoid and find distressing those situations that have been highly disadvantageous in species survival. Disruption of inter-individual bonds in primates is perceived as seriously threatening. It appears unpleasant and is associated with physiological responses of alarm and mobilization. Such disruptive events usually stimulate coping behaviour that tends toward restoration of attachments. This formulation is consistent with field observations and laboratory experiments involving higher primates.[17] When separation occurs and obstacles interfere with reunion, intense and persistent efforts are made by both mother and offspring, directed towards restoring contact. If these efforts fail, for whatever reason, they are likely to be followed by behaviour that bears

Ill. 33. The grief-stricken faces of bereaved parents of earthquake victims show the physical manifestation of a vital behavioural pattern. The powerful, enduring attachments between parents and offspring have been crucial in human survival.

Ills. 136, 137, 140

Ill. 33

considerable similarity to human depression.[31] These observations of reaction to inter-animal loss in the natural habitat are consistent with laboratory investigations of a separation-induced model of depressive behaviour in monkeys.[23] In these experiments, a mother and infant are separated, or juveniles who have been reared together are separated. Thus a strong bond is disrupted. The main tendency is towards a depression-like response in both mother and infant. In the infants, typically there is initial distress: calling and searching for several days. This is followed by greatly diminished activity, decreased play, huddling posture, and reduced food intake. They resemble humans who report feeling depressed. There may be subtle lasting effects due to brief separation.[24]

Is there anything comparable to this depressive syndrome induced by separation or loss in humans? Clinicians have made observations of personal loss, grief, and depression.[35] Patients often come to medical attention in the context of an important loss. The most vivid circumstance is the grief reaction to the loss of a personally significant individual. The grief reaction is a specific pattern of distress in which the person's focus is on the loss. Usually, gradual recovery occurs through a difficult process of mourning over a period of months. However, some persons become clinically depressed, experiencing a pervasive undermining of prior interests and human relationships, with feelings of despondency.

In the past few years, several research groups have undertaken the systematic study of possible relations between life events and the onset of depressive episodes serious enough to come to psychiatric attention.[27] Overall, the present evidence indicates that the experience of interpersonal loss elicits much distress—grief is one of the most difficult transitions of the life cycle—and it is a common precipitating factor in clinical depressions. However, such depressions need not be triggered by separation or loss and, equally, separation or loss may trigger distress reactions other than depression.[25] Indeed, most occurrences of separation and loss do not precipitate clinical depression even though a period of grief or at least sadness regularly occurs. A few studies have explored ways of coping with separation and loss.[22] The variety and effectiveness of such patterns in the general population is impressive. These studies highlight the question of differential susceptibility to loss and grief. What is the special vulnerability of those who become overwhelmed with despondency? Clarification of this issue is largely a matter for future research,[23] but the general link between jeopardy to close human relationship and emotional distress is well established.[40.5]

Thus, there has in all likelihood been an evolutionary premium on the capacity to form attachments because of their adaptive utility in the line that led to *Homo sapiens*. Human groups can usefully be viewed as pools of potential attachments. This attachment-pool manifested itself in somewhat different ways in different evolutionary eras, but always there was a small core group of well-known individuals, some of whom were ready at a moment's notice to be helpful. These considerations apply to: (1) the millions of years in which some non-human primates were slowly evolving in the hominid direction; (2) the several million years in which hominids were organized in hunting-and-gathering societies of about fifty people, extended somewhat by kinship relations with neighbouring bands;[32] (3) the extended family of agricultural village society;[33] (4) the primary group of the homogeneous neighbourhood in pre-industrial towns.[42]

These small groups shaped the world of human perceptions, values and beliefs. They provided the security of thorough familiarity, support in times of stress, clear-cut guidelines for behaviour, and enduring attachments through the life cycle. These are powerful assets, strongly valued by humans everywhere, resistant to erosion. Yet other powerful values have indeed modified these relations in the twentieth century in a way that must be largely unprecedented in the long course of human evolution.

The advent of high technology has made possible a dramatic increase in man's control over his environment, including some aspects of great historical significance: reduction in mortality and increase in life expectancy; and increase in food supplies, resources and gratifications of many kinds. To take advantage of these prized opportunities, it has been expedient to increase geographical and social mobility and to foster very large concentrations of people in urban areas. Nowhere is this process clearer than in the U.S.A., though it is certainly not limited to any one country. Today, individuals or a very small nuclear family move freely from one place to another, often entering a community where they initially have no personal acquaintances at all. In the process, they are likely to attenuate or disrupt attachments that have been meaningful in their lives. This increases vulnerability to separation and loss and poses a threat to the sense of belonging in a valued group. It may be a factor in the pervasive problems of sadness, grief and depression.

There is a paradox here. The disruption of close attachments so common in highly mobile societies occurs in the context of a larger group of superficial acquaintances and impersonal contacts than ever before in history. In this world of huge cities and mass travel, there is no dearth of human contact, especially with those who are more or less strangers. But the smaller group who are intimately known and who can be counted upon to care and help without question is commonly only one or two—or even zero.

Ill. 34. Opposite: the impersonality of life in a large city is underlined by the seething mass of Christmas shoppers in London. This scene illustrates the paradox of the disruption of inter-personal attachments in highly mobile societies; even in the middle of a crowd, many people who are uprooted feel lost and alone.

Ill. 34

POPULATION

Earlier in this chapter, I have sketched the enormous population growth that has occurred within a moment of recent evolutionary time. I now wish to consider briefly some of the factors in biological and cultural evolution that tend to keep this powerful current flowing as it is, even today, throughout most of the world.

The emotions of sexuality are part of a strong motivational pattern that has met a long-standing need of the species to produce offspring. Throughout most of human evolution, the production of offspring had to offset a high infant mortality rate. But with the recent advent of remarkable adaptive advances—effective food production, sanitation, public health and medical care—infant mortality and other pre-reproductive mortality have diminished, although the motivation for copulation has not.

The human species has evolved rather strong sexual motivation, manifested, for example, by frequencies of copulation that are high among the primates. Moreover, male and female maintain sexual interest in each other throughout the menstrual cycle, rather than being restricted to an oestrus portion. Thus, the likelihood of conception by women in their reproductive years has historically been high, and conception can occur again relatively soon after each pregnancy. Moreover, there is a very widespread cultural emphasis on strong emotional investment in each infant and a high premium *Ill. 35* on care of offspring who survive infancy. This orientation appears to be learnt with ease and may well be based on a biological predisposition built into the brain because of its effectiveness in adaptation; consider in this context the profound tenderness so readily elicited in many humans by the sight of a tiny infant. Yet these human commitments in sexual and parental behaviour were, for many thousands of years, swimming upstream against a powerful current—high infant mortality from disease, injury, and perhaps predation; moreover, periodic food restrictions probably exerted a strong restraining influence on population size. Some scholars estimate that, at the onset of agriculture (not very long ago by evolutionary standards), there were more baboons and macaques on earth than humans. First agriculture, then expanding science and technology—especially in sanitation and health—lifted the pressures that had held down human population growth for something like ninety-nine per cent of our history as a species. With the pressure lifted, the dramatic population growth described earlier surged forward on a scale dwarfing all previous human experience. The consequences of this growth are now having a powerful impact on nearly all the countries of the world—north and south, east and west, rich and poor. Brown[6] has called attention to some important aspects of this situation:

> During the nineteenth and early twentieth centuries, technology was transplanted to the poorer countries in a one-sided manner, which resulted in tremendous dislocations. Lifesaving technology reduced death rates, and with birth rates unchanged, populations increased rapidly. At the same time, modern agricultural and industrial technologies were introduced to these areas much more slowly, with the result that the lot of the average inhabitant became increasingly miserable. . . . By contrast, in the *Ill. 30* richer countries, technological change and agricultural and industrial production are increasing at rates that are much greater than the rates of

Ill. 35. The strong emotional investment in each human child, visible in
the tenderness elicited by the sight
of babies, may well be based on a
biological predisposition because of
its effectiveness in survival.

population growth. . . . They must consume vast quantities of raw
materials and energy of decreasing availability in order to produce the
goods that people want, and in the process their factories excrete equally
vast quantities of waste products into rivers, lakes, oceans, and the
atmosphere. Technological man is changing his environment with unprecedented speed, and the ultimate effect of those changes can be only
dimly perceived.

Ill. 36

Various scholars have recently reviewed the adverse consequences of rapid
population growth on the current unprecedented scale.[11,41] These considerations include effects on economic development, health, education,
pollution, resource depletion, conflict and violence. Since these risks are
becoming increasingly clear on a world-wide basis, why do they not lead to
major adjustments? Since undesirable side-effects of technological advance
may be countered with other technological advances—i.e., a technological
fix—why not solve the problem in this way? Indeed, the present state of
contraceptive technology is impressive, so why not apply it effectively before
the problem gets out of hand? It turns out that old attitudes and values
that have served mankind well for a long time now interfere with a flexible
response to the new conditions of life. Davis[7] summarizes the situation:

> Human societies have always had difficulty in motivating people to pursue
> goals that benefited them collectively rather than individually. In the past,
> they ingeniously solved the reproductive problem by building into insti-
> tutional structures powerful incentive systems that yielded high birth
> rates in the face of calamities, hardships, and perpetual poverty. Now—

suddenly, in terms of biological history—societies must do an about-face. The past response to good conditions has been sizeable families and population growth. Human beings are now being asked to respond differently.

Indeed, the utilization of effective contraceptive technology has on the whole been slow and faltering, even when it has been available at reasonable cost and encouraged by government policy. Why is this so? No doubt many factors influence the long-standing human preference for large families; here are a few that deserve consideration: (1) the ubiquitous security of a primary group, committed fundamentally to mutual support and assistance; (2) the economic and psychological insurance value of a large family in protecting against loss by death or migration; (3) the tradition in many agricultural societies of landholding being transmitted through sons; (4) the widespread historical belief that increasing group size will yield political and/or military advantages—perhaps even linked to the ancient primate experience of domination of small groups by large groups.

In other words, customs and perhaps biologically-based orientations that worked well for a long time in earlier environments tend to persist in the

Ill. 36. Pollution. One of the most vivid ways in which man is changing his environment at a constantly increasing rate is the thoughtless way in which he is not only using the earth's resources, but also wasting and irreversibly destroying them, thus upsetting the balance of nature.

Ill. 37

face of new environmental conditions. Their persistence is enhanced by the fact that they have long been believed to be useful in meeting the problems of living and so they have been invested with strong emotions and taught early in life—when long-term commitments are most likely to be established. Thus, learning new patterns of behaviour to meet unprecedented circumstances is difficult—though surely not impossible.

AGGRESSION

The ubiquitous and dangerous character of human aggressiveness raises serious questions. What is the evolutionary history of such behaviour? What present lines of enquiry offer some promise of clarifying its nature? Do modern societies impinge on man's biological nature in such a way as to heighten aggressive tendencies? What recent changes in environmental conditions are especially pertinent to hostility and violence?

Aggressive behaviour has been an important part of adaptation in higher primates for millions of years. Patterns of threat and attack have probably been useful in several ways: increasing the means of defence; providing access to valued resources such as food, water, and females in reproductive condition; contributing to effective utilization of the habitat by distributing animals in relation to available resources; resolving serious disputes within the group; providing a predictable social environment by establishing status relations; providing leadership for the group, particularly in dangerous circumstances.

Threat and attack patterns common to a variety of primate species, such as chimpanzees, gorillas, and baboons, are similar to some of the aggressive patterns of our own species. But what are the conditions under which aggressive patterns are likely to occur in higher primates? The principal ones so far noted are as follows: (1) in daily dominance transactions; (2) in redirection of aggression downwards in the dominance hierarchy; (3) in the protection of infants; (4) when sought-after resources, such as food, or sexually receptive females, are in short supply; (5) when meeting unfamiliar animals; (6) in defending against predators; (7) in the killing and eating of young animals of other species; (8) when terminating severe disputes among subordinate animals; (9) in the exploration of strange or dangerous areas; (10) when long-term changes in dominance status occur, especially among males; (11) when an animal has a painful injury; (12) when there is crowding of strangers in the presence of valued resources.

Thus, it is not unreasonable to suggest that there are similarities in the forms and contexts of aggressive behaviour between higher non-human primates and those of *Homo sapiens*.[30] The higher primates are particularly likely to behave aggressively in relation to *defence* and *access to valued resources*.[18] *Crowding of strangers* intensifies these aggressive responses.[19] Behaving aggressively in the contexts of defence and of access to valued resources may well have given selective advantage in the evolution of higher primates. If adequately regulated, such behaviour can serve to enforce or implement a variety of adaptive needs, such as those involving food, water, predation, and reproduction.

Natural selection probably operated in primate evolution to delineate modes of regulating aggression and resolving conflict. Primates have clear-cut patterns that are usually effective in terminating aggressive sequences.

Ill. 37. An example of the persistence of biologically based orientations, which worked well in earlier environments, even in the face of new conditions. This anti-abortion rally underlines the participants' belief in the sanctity of life, despite the now urgent need to decrease the current rate of population growth.

43

They display an elaborate repertoire of submissive behaviour. They have stable dominance relations that contribute to the predictability of the social environment. The chimpanzees have interactional sequences of aggression–submission–reassurance that have elements in common with the behaviour of humans in similar circumstances. Altogether, it is likely that man has a vertebrate–mammalian–primate heritage of aggressive tendencies which has been transmitted both biologically and culturally. In order to understand the nature of human behaviour, it is important to give attention to the sources and instigation of aggression as passed on in evolutionary history, and to look at the biological mechanisms of control and regulation. Any evolutionary heritage of aggressive tendencies must be mediated through genetic influences on brain mechanisms and hormones that affect the brain. These mechanisms must be highly dependent upon environmental factors, especially in the social environment, for the course that their development takes in the individual life cycle.

The human species has a biological heritage of its vertebrate–mammalian–primate history.[49] This clearly includes some features of brain and behaviour.[45] Such evolutionary relations are highlighted by the recently discovered biochemical and immunological similarities of chimpanzee and

Ill. 38. Whether or not this beautiful painting of a horse fulfilled an aesthetic desire on the part of the unknown cave artist, it was created for a very specific ritual purpose. The lines above the horse's back have been interpreted as arrows, and it would seem that the painting either symbolized a future successful hunt, or depicted one that had already taken place. In all likelihood, it was intended to enhance the capacity of the group to meet survival needs.

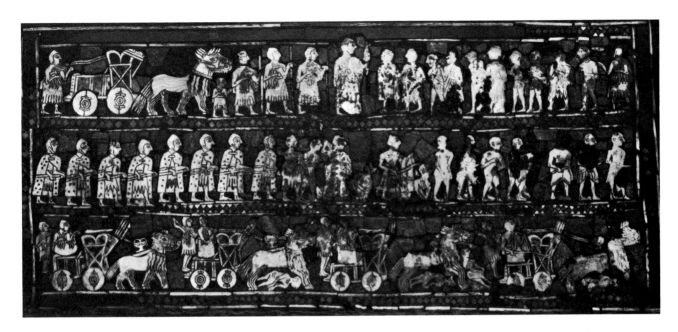

Ill. 39. Violence has been a part of man's way of life for millennia. The 'war' side of the Royal Standard of Ur, dating from *c.* 2700 BC, illustrates vividly the aggressive behaviour of the Sumerians towards the enemy and the lack of mercy shown to their prisoners.

human. The structure, organization, and circuitry of the chimpanzee brain has remarkable similarities to that of the human. It is perhaps not surprising that a brain so similar to ours should produce behaviour, including aggressive behaviour, similar to our own. These similarities in basic elements of aggressive behaviour include: the forms of threat, attack, submission, and reassurance; the contexts and eliciting conditions of aggressive behaviour; the role of co-operation in facilitating aggression; the tendency to make status differentiations; the tendency to redirect aggression towards weaker or lower-status individuals; the adolescent male aggression spurt; the hostility between relative strangers belonging to different communities. These similarities occur at the level of individual and small-group behaviour. So far, there has been no observation of behaviour remotely similar to human warfare.

Ill. 39
Ill. 38

Whatever the evolutionary background and its biological legacy, the historical record makes it clear that aggressive behaviour between man and man, between man and animals, and between human groups has been a prominent feature of human experience for a very long time. Such behaviour has been easily learnt, practised in play, encouraged by custom, and rewarded by most human societies for thousands of years. Evidently these patterns are learnt and are a part of cultural traditions. Yet they seem to be learnt with great ease and facility in a wide diversity of cultures. Perhaps the human organism is primed to acquire certain elementary behaviour patterns with relative ease. There may well be a special facility for learning in directions that have been adaptively valuable for the species over a very long time in evolution.[16] For any species, some patterns of behaviour are easy to learn, some difficult, and some impossible. Learning in such adaptively-significant spheres as those concerned with food, water, and reproduction has probably had high biological priority; and aggression can serve in the implementation of these adaptive requirements. So the human brain may well be organized in a way that reflects the long-term selective advantage of facility in learning

45

such behaviour.[17] Simple preferences on the part of the infant or young child might draw his attention to a certain class of stimuli, or reward his engagement in a particular kind of activity—for example, rough-and-tumble play.[21] Once drawn in this direction early in life by an inherited preference, a great deal of complex learning could ensue, taking full account of cultural instructions.

The biological equipment of the human organism is surely very old. Some of our emotional response tendencies and learning orientations, mediated by the old limbic-hypothalamic-midbrain circuits, were probably built into the machine because they worked well in its adaptation over many thousands or even millions of years.[37] There has been very little time for change in that structure since the Industrial Revolution began two centuries ago. Yet, the circumstances of our present life are largely a product of that revolution.

The adaptability of aggressive behaviour in past environments gives no assurance whatever that similar behaviour would function adaptively in the very different environment of contemporary man. Scholars differ as to the fate of primate aggressive tendencies during the two to three million years that man has lived in hunting-and-gathering societies; but there is reason to believe that tendencies of this sort were strengthened by the advent of agriculture and permanent settlements 8000 to 10,000 years ago. Such heightening of aggressiveness may only be a matter of social expression, but it may also reflect changing selection pressures that modified the basic nature of the human organism. However that may be, the rich historical record is replete with suspicion, hatred, and violence; and the advent of complex weapons technology has suddenly and drastically escalated the risks of such behaviour.

Ill. 40

One consistent finding from recent studies of complex non-human primates in natural habitats, semi-natural settings, and laboratory experiments meshes with earlier work in a variety of non-primate species. There is an ancient vertebrate–mammalian–primate tendency to respond aggressively when strangers are crowded together in the presence of valued resources.[19] There are some indications, especially from the evolution of behaviour in cities, that this tendency has been maintained and perhaps even accentuated within recent times.

Since the contemporary environment is increasingly an urban environment, the history of cities deserves examination in relation to aggressive behaviour. Sjoberg[42] indicates that the pre-industrial city was already crowding strangers beyond most prior conditions. The buildings were crowded together and there was a high density within them. These cities were characterized by strict social segregation, exemplified in sharply delineated 'quarters'. Often these quarters were separated from one another by high walls and were locked at night. Each ethnic group lived separately and utilized a distinct occupational niche. High-status groups usually lived in the centre of the city while low-status groups lived on the periphery.

Ill. 41

Each group provided a cohesive social unit with clear guidelines for behaviour among its members, and extensive opportunity for children to observe the activities of adults. Bonds among the members of each group were strongly reinforced through common religious beliefs and activities. Thus, the residents of one quarter of a medieval city tended to be relative strangers to those in another quarter. The pre-industrial city fostered sharp

Ill. 40. Man's inhumanity to man continues unabated in the twentieth century. This anti-American poster dating from the Vietnam war forcefully depicts the ruthlessness and hatred of the aggressor.

46

in-group, out-group distinctions, and the tensions typically associated with such distinctions. One may wonder what legacy in outlook and attitude these cities may have left us.

For several thousand years, such pre-industrial cities slowly evolved in the context of a predominantly agricultural society. Then the Industrial Revolution occurred and the human population rapidly became urbanized in *Ill. 42* industrialized countries. From an evolutionary viewpoint, these changes are very recent, they have moved with extraordinary speed, and are far from completed. In 1900, only one nation was urbanized (Great Britain), i.e., a majority of the population lived in towns. In a few generations, with little if any time for change in man's biological equipment—and not even much time for change in customs—a drastic transformation has occurred in man's environment.[38] The speed and magnitude of this environmental transformation has been so great, and so unprecedented, that it would be miraculous if effective adaptation to the new conditions had been achieved by now.

Cities crowd strangers beyond anything known in the past, involving: *Ill. 34* vast numbers of persons; mobility, which brings a continuous flow of strangers into the city; and complex living, which brings people regularly into contact with many strangers, most of whom are never contacted again. This crowding of strangers occurs in the presence of a richer array of valued resources than ever known before, often perceived as being in short supply: valued objects, activities, persons, and places.

These circumstances foster anonymity, impersonality, and harshness in human relationships.[19] Violence becomes more common. The sense of mutual responsibility tends to be diminished. Some societies are more successful than others in coping with these problems, and many factors influence the outcome.[36] But the problems are sufficiently serious and widespread that

Ills. 41, 42. The symmetry of pre-industrial Georgian Edinburgh compared with the medieval 'old' city (left) typifies the Neoclassical intellectual approach to civilization. Piranesi's *Carcere* (opposite), though dating from the same period and inspired by the same ideals, seems to look ahead to the de-humanizing effect of the 'dark satanic mills' of the Industrial Revolution.

they deserve frank recognition and greater attention from the scientific community—in the perspective of their recent arrival in terms of human evolution. On the time-scale of man's origin, we have little experience in coping with such problems. Habitual ways of approaching them may well be inadequate. For instance, one of the customary ways of responding to the stresses of city life is for individuals to identify strongly with an ethnic or

Ill. 43 political group and seek improvement of conditions through group effort, thus going beyond what any individual, no matter how capable, might do. Then, groups come into conflict, competing for more or less scarce resources. By this, I mean not only economic but psychological resources—recognition, credit, esteem, status, respect. A common but potentially lethal aspect of such inter-group conflicts is the tendency to blame other groups for the dissatisfactions of one's own and ultimately (though not necessarily) to

Ill. 44 justify great harm to them in terms of such blame. With regard to both inter-group and inter-individual conflict, one may wonder whether we are a blaming species.

In contemporary societies, conflict between groups is common, sometimes quite destructive, and highly varied in content. Yet there are widely

49

shared properties in the form of such antagonisms that may help to clarify some relations between social stress and aggressive behaviour. Human societies have a pervasive tendency to make distinctions between good and bad people, between heroes and villains, between in-groups and out-groups. This sorting tendency is very widespread, readily learnt, and quite susceptible to harsh dichotomizing between positively valued 'we' and negatively valued 'they'.

Hostility between human groups is likely to arise when the groups perceive a conflict of interest, an unacceptable difference in status, or a difference of beliefs that jeopardizes self-esteem. Such situations tend to evoke sharp in-group, out-group distinctions, with drastic depreciation of the out-group by the in-group. Perceived threat from an out-group tends to enhance in-group solidarity, tightness of group boundaries, and punishment of those who deviate from group norms.

Justification for harming out-group members rests on sharp distinctions between 'we' and 'they', between good people and bad people. Such justification is readily provided by assumptions regarding: (1) the damage they would do to the in-group; (2) the damage they would inadvertently do to themselves; (3) classification of the out-group as essentially non-human.

A great variety of political, social, economic, and pseudo-scientific ideologies may be mobilized in support of these hostile positions. Although the content of such inter-group hostility varies widely from time to time and from place to place, the form of the antagonism is remarkably similar.

This group-centred view of life has been studied by social scientists and is usually referred to as 'ethnocentrism'.[34] It may be thought of as an extension of 'egocentrism'. It seems easy for most of us to put ourselves at the centre of the universe, attaching a strong positive value to one's self and one's group, while attaching a more or less negative value to many other people and their groups. At the extreme, the situation often falls into the pattern described by LeVine & Campbell: 'Symbols of one's own ethnic or national group or of the values shared by that group (or both) become objects of attachment, pride and veneration; symbols of other groups or their values become objects of contempt and hatred.'

Since these patterns are not limited to ethnic groups, a term like 'groupocentrism' would be more appropriate. In studies of 'ethnocentrism', groups have in fact been specified not only by ethnicity but by religion, race, language, region, tribe, nation, and various political entities. The same principles seem to apply across these various groups. A crucial question is whether groups can achieve internal cohesion and self-respect without promoting hatred and violence. A deeper understanding of factors that exacerbate and ameliorate groupocentrism could have great practical value in resolving inter-group conflicts in the future. This is, in my view, one of the most important areas for scientific enquiry. Valuable contributions have been made.[8,10,13,43] But here, too, the changes in the human predicament have been so rapid that traditional responses have been inadequate. In this instance, policy-makers and scientists have not, until very recently, thought of these problems as suitable for scientific study, and a miniscule portion of scientific effort has moved in these directions.

Contemporary inter-group conflict is affected by two other recent developments of cultural evolution: the prevalence of culturally heterogeneous

populations within nations, and the limited experience with democratic forms of government. Revelle[41] points out such difficulties in heterogeneous, rapidly-growing, developing countries. Similar considerations apply to many industrialized countries:

> ... rapid population growth creates or aggravates political and economic conflicts between racial, cultural, religious, and linguistic groups. Numbers are an important element of political power, especially in countries that are attempting to introduce or to maintain democratic institutions and processes. ... Conflicting groups usually perceive these changes to be much larger than they really are. ... Since income and occupation often coincide with differences in religion, caste, tribe, or language, a half-developed system of primary and secondary school education may sharpen both class and ethnic differences. ... Inter-group relations may be further exacerbated by another result of rapid population growth in most less-developed countries—large-scale migration from the country to the city or from one region to another. ... In some cases, far-reaching measures, such as mass migration or fragmentation of states into autonomous or semi-autonomous smaller units, may be the only feasible policy options. But governments can do much by a more even-handed treatment of different groups.

Human individuals and groups are clearly capable of being considerate to others, kind and helpful, even generous. Conflicts are ubiquitous, but usually resolved short of hatred and violence—sometimes creatively with lasting benefit. Yet the problems of enduring conflict, hatred and violence are clearly visible and involve higher risks than ever before—if for no other reason than the destructive capacity of high-technology weapons, their widespread availability, and the easy justifications for violence in many diverse cultures

Ills. 43, 44. Strong identification with a political or ethnic group is one of the ways in which people respond to urban life. In the photograph (opposite) of a manifestation for and against the European Economic Community, supporters of each side seem to be addressing their demonstration to a third party, rather than confronting each other. In the photograph below, the confrontation has become more violent and the pervasive distinction between 'good' people and 'bad' people is strengthened by the juxtaposition of the bayonets and the flower. The tendency of human groups to depreciate each other harshly, and to justify violence towards those so depreciated, is exceedingly widespread and dangerous.

around the crowded planet. Therefore, it becomes important to understand the conditions under which humans are likely to injure one another, and the roots these conditions may have in the tortuous road of man's origins.

What frustrations are likely to elicit severe aggressiveness in contemporary humans? One crucial kind of frustration is the clear and present danger to survival. The evolutionary roots of this response tendency are straightforward. In non-human primates in their natural habitats, recent research has shown that, for survival, it is useful to be an integrated member of a social group; moreover, anthropological studies highlight the crucial role of group membership in the evolution of early man. For millions of years, man's survival has been greatly facilitated by attachment to other humans and by belonging to organized groups orientated towards adaptation.

Frustration of self-esteem or sense of personal worth also has a strong tendency to elicit hostile responses. The need for self-esteem is related in evolution to the adaptive value of having a place in the group. Over long time-spans, an individual's chance of survival and reproduction must have been facilitated by being respected by others. Such respect was enhanced by established competence appropriate to one's age or sex class. Behaviour directed towards establishment and maintenance of self-esteem, of dignity, of personal worth is expressed in many ways in diverse cultures. It appears to be a basic human need.[14] Historically, hunting and warlike activities have been an important source of self-esteem for males. In the contemporary world, less destructive bases for self-esteem will have to be found.

Frustration in crucial inter-personal relationships is also likely to elicit hostile responses. These are the primary relationships for group membership and are highly significant in the evolution of our species, as indicated above. Closely related is a threat to the sense of belonging to a larger group, beyond the intimate few, but a group with which one identifies. It may be an ethnic group, a nation, a tribe, a political entity, or an occupational unit. Such groups make a contribution to the self-esteem of their members and provide guidelines for adaptation.

These basic motivations, then, have deep roots in the soil of human origins. When these motives are frustrated, the likelihood of aggression is increased. Threats to these fundamental motivations may also elicit ways of coping that are not hateful or violent. Serious frustrations may lead to assertive behaviour, personal initiative, vigorous and persistent efforts towards constructive problem-solving. This course may be more difficult, complicated, and tedious in the short run but much more rewarding in the long run. The poignant dilemma is that ways of fostering survival, self-esteem, close human relationships, and meaningful group membership for hundreds, thousands, or even millions of years now often turn out to be ineffective or even dangerous in the new world which man has suddenly created. Some of the old ways are still useful, others are not. They will have to be sorted out, and sorted out soon. Where old ways are no longer suitable, new ways must be found—not valued because they are new but because careful observation demonstrates their suitability for the changed conditions.

The small, slowly-changing world of the non-human primates and of early man is gone now—and probably gone forever. So, too, is the fairly small and simple world of agrarian society. Our world is the new, large, crowded, heterogeneous, rapidly changing world of the past few centuries.

Ill. 45. It has been said that the cardinal virtue of this generation is neither faith nor love, but hope.

There is little in our history as a species to prepare us for this world we have made. It is, for instance, difficult for us to identify with, feel for, or care about many different people—especially those who are far away, or who look different from ourselves, or who have different customs.

Will it be possible for us to achieve unity-in-diversity? Can the rich variability of human biology and human behaviour be appreciated and preserved in a single world-wide community? Can ways be found to meet human needs equitably, to respect each other and our habitat, to resolve conflicts before we destroy ourselves and our environment?

Efforts to mobilize the strengths of science for this task are now under way, but are so far at an early stage of development.[3,6,11] Attitudes, emotions, beliefs, customs, and political ideologies from our past will often hinder the utilization of scientific knowledge even when it is available. But our motivation for survival is strong, our problem-solving capacities are great, and the time is not yet too late.

To adapt to these unprecedented conditions of the modern revolution-in-evolution, we must understand more deeply than ever before the nature of the human organism, the forces that shaped the organism in the very long course of its evolution, the drastic transition of this moment, and the ways in which the new conditions impact on our old apparatus and its habitual modes of adaptation.

If there is any agenda more important for science and society, I wonder what it could be?

Ill. 46. The United Nations, although beset by grave political problems, provides a meeting-place for the enormous diversity of human societies. Whether it can help to overcome traditional fears and hatreds remains to be seen.

Ill. 46

The Structure and Cultural Development of Man

3 The Tree of Evolution

John Napier

MAN possesses a constellation of physical and behavioural characteristics, some of which are uniquely his own, some of which he shares with his nearest primate relatives, others which are characteristics of the primate order as a whole, and still others which are the property of all living creatures.

In defining man, a necessary preliminary to searching for his remains in ancient rocks, the problem is to determine which of his many characteristics are uniquely human—what in fact are the hallmarks of mankind?

Characteristics, or rather characters, as they will be called from now on, can be divided arbitrarily into three kinds: structure, function and behaviour.

In the language of automobiles, *structure* is represented by the component parts, *function* is the integrated action of these parts when the engine is started, the gears engaged and the car is driven off the assembly line, and *behaviour* is the performance of the car on the highway, which is its natural environment. Although each set of characters are independent of one another in one sense, they are all part of a single whole in another. There can be no behaviour without function and no function without structure.

The success of an organism ultimately depends on its behaviour: there is no virtue in a car that does not hold the road or whose petrol consumption is impossibly extravagant. Competition from other cars will soon drive it into obsolescence. In nature the same will happen to an animal that is incapable—through faulty design—of reproducing its own kind. This, simplistically, is the basis of the theory of survival of the species. It is not surprising that natural selection, which is the mechanism of evolution, operates at the level of behaviour. In the car analogy, natural selection is replaced by consumer choice: no one is going to buy a car that looks good and sounds good when ticking over but is incapable of holding the road.

However, although theoretically behaviour is the ideal hallmark, we are forced to settle for something less spectacular in the quest for the milestones of human evolution. One can't dig up behaviour, nor function for that matter, but one can excavate structure.

Thus, inevitably, a quest for the origins of man is, in its earliest phases, rather a prosaic affair, an activity for practical people like field geologists who can locate a fossiliferous site, zoologists who can distinguish between human and animal remains, technicians who can extract delicate bone fragments from a hard, stony matrix and reconstruct the pieces into an acceptable whole, anatomists who, by means of simple measurements and complex statistical strategies, can interpret the topography of bones, and physical chemists who can put some kind of a dateline on the relic.

Ill. 47. Opposite: a Neanderthal skull from Shanidar, Iraq, discovered by Professor Ralph Solecki. The skull, which is only partially developed from the matrix, is about 45,000 years old. Observe the prominent brow ridges and large, flattened skull vault.

In their wake come the theorists, the interpreters, who expand the structural evidence, clothe it with flesh, animate it with function, and even sometimes infuse it with a modicum of behaviour.

Another group of scientists who enter the lists at this stage are the taxonomists. Their interests are in classification of animals (human or otherwise). They are also concerned with supplying names for animals in accordance with the binomial principle laid down by the Swedish naturalist Linnaeus and 'policed' nowadays by an international watch committee.

Classification and naming, oddly enough, are the most contentious elements in the inauguration of a new fossil species. More heat has been engendered, more tempers frayed and more lifelong friendships dissolved by taxonomic disagreements than by any other form of anthropological dispute. But perhaps this is all to the good. Already there is a noticeable caution abroad and the somewhat irresponsible latin-naming of every 'new' species that comes into the hands of the field worker, so prevalent in the past, is being replaced by the sensible practice of identifying the specimen by a non-committal field number or a geographical location. For example, Richard Leakey's recent discovery, an exciting new human skull, is known as KNM-ER 1470 or the East Rudolf skull (see p. 99).

Ill. 94

For all its routine and technical procedures, the quest for the origins of man is an intensely romantic odyssey that cannot simply be dismissed as workaday routine. The desire to know more about himself is surely one of man's strongest motivations and lies at the very roots of his social and cultural attainments. The command, 'know thyself', was emblazoned in gold letters over the portico of the temple at Delphi—a precept that, according to Juvenal, descended from heaven. To go from the sublime to the ridiculous, the homespun philosophy of many young people today to do 'one's own thing' is simply the contemporary way of complying with the Delphic instruction.

Man's search for self-knowledge has many forms. In allegory, the Golden Fleece and the Holy Grail were quests of such a nature; and in historical times, the great journeys of geographical discovery and the epic adventures that accompanied the burgeoning of the scientific method kept the pot boiling. Today, in the form of polar journeys, mountain climbing, ocean adventures of one sort and another and, above all, in the exploration of outer space, the physical side of man's search for the truth about himself and his environment is being well catered for. But of greater importance, perhaps, is the flowering of a more intellectually orientated quest, that directed towards achieving a greater understanding of the nature of human attitudes, the roots of our behaviour and the foundations of human societies.

The central issue of anthropology today pivots around a single question: is man's behaviour simply a matter of tradition and history, something he learns first at his mother's knee, and later in the groves of Academe and the alleyways of common experience? Or is it a heritable commodity transmitted through the genes reflecting the accumulated behavioural adaptations of his distant past? This obviously raises two further questions. What is the nature of man's past and how can it be investigated?

We have already agreed that the study of fossil man cannot supply any *direct* evidence about the behaviour of prehuman and early human populations. Thus, it would appear that, failing fossils, the only alternative is to investigate living man, a line of enquiry that is fraught with difficulties. *Homo sapiens* has advanced too far and too fast and his basic patterns of behaviour are buried forever beneath the cultural avalanche that has swept over him in the process. That seems to rule him out as a source of comparison, but what of the 'lost' or 'primitive' tribes like the Hadzas, the Bushmen, the Tasadays of the Phillipines? Are they not still, technically, in the Stone Age? The Kalahari Bushmen have been the subject of many anatomical studies but relatively little is known about their day-to-day behaviour (but see Lee and DeVore's recent book, *Man the Hunter*, and Eibl-Eibesfeldt, below, Chapter 7). Although the Bushmen have been long isolated from the mainstream of human cultural evolution, it must be assumed that they also have evolved their own particular behavioural specializations, in addition, of course, to myth, language and the ability to reason—facilities common to all races of living man. Apart from a certain amount of ecologically orientated behaviours such as hunting and gathering, they are just as unsuitable as models for protohominid behaviour as are the rest of us.

However, there is still one further strategy to be considered. If extrapolation backwards from 'primitive' *Homo sapiens* fails to be a meaningful approach, then perhaps extrapolation forwards from the apes and monkeys will supply the answers we need about the way of life of primitive man and, thus, about the foundations of our own mores and morals.

Forward projection from the apes and monkeys is a popular sport but one, nevertheless, that remains a constant source of controversy. Fuel has been added to the flames by such able writers as Konrad Lorenz, Robert Ardrey, Desmond Morris, Robin Fox and Lionel Tiger, and the fire still burns brightly. This is not the place to discuss at length the merits or otherwise of this approach, which depends on information derived from field studies of living primates, except to observe that—to date—field studies have taught us more about primate behaviour than about human behaviour!

Ill. 48   Exactly how much of this fascinating information, which is of immense importance in its own right, can be deployed to further our knowledge of early man depends on a fuller understanding of the zoological place of our human ancestors within the order *Primates*. Clearly, the longer the time-lapse since the ancestors of living apes and men diverged from a common stock (a zoological relationship that is generally accepted), the greater has been the opportunity for independently acquired characters to have evolved in both descendant families. It can be argued that since the differences between the two families have become more pronounced with time, the longer the period that has passed since their separation, the less meaningful are deductions based on the observed behaviour of living apes, and vice versa. Only the continued excavation of fossil evidence can supply us with the answers to such questions as when did the apes-to-be and humans-to-be part company and go their separate evolutionary ways? Was it twenty million years ago, or ten or five? These are the vital, if prosaic, questions.

The above argument supposes that structural affinity between two groups of animals depends on homology or genetic continuity, but there is another basis for similarity which is independent of close zoological affinity. Analogous structures may develop in two unrelated groups of animals which occupy the same or similar habitats. The more closely related the animals, the more likely it is that the analogy will be close. For example, there are

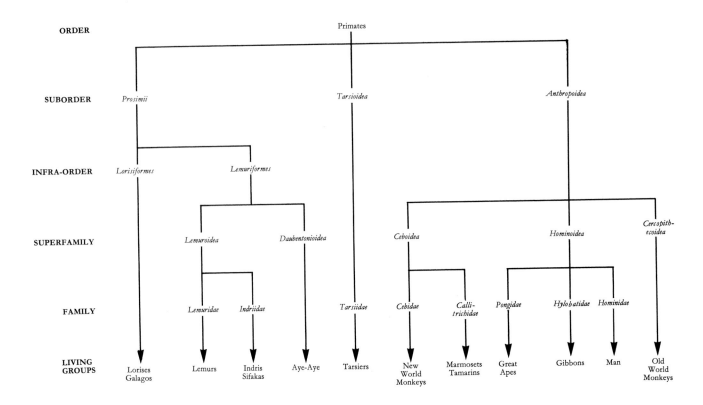

ORDER					Primates							
SUBORDER		*Prosimii*			*Tarsioidea*			*Anthropoidea*				
INFRA-ORDER		*Lorisiformes*	*Lemuriformes*									
SUPERFAMILY			*Lemuroidea*	*Daubentonioidea*			*Ceboidea*		*Hominoidea*			*Cercopith-ecoidea*
FAMILY			*Lemuridae* *Indriidae*		*Tarsiidae*		*Cebidae*	*Calli-trichidae*	*Pongidae*	*Hylobatidae* *Hominidae*		
LIVING GROUPS		Lorises Galagos	Lemurs Indris Sifakas	Aye-Aye	Tarsiers		New World Monkeys	Marmosets Tamarins	Great Apes	Gibbons Man		Old World Monkeys

broad similarities between certain marsupials (mammals with pouches) and eutherians (mammals with placentas), although their evolutionary histories are as disparate as they could be. There are marsupial *doppelgangers* for such eutherian creatures as dogs, cats, bears, rabbits and moles, but nobody would suggest that they are very closely similar in appearance or that you could mistake one for the other; it is simply that they occupy and are adapted for similar environmental niches. Among eutherians, particularly eutherians of the same zoological order, physical similarity is striking. Amongst primates, for instance, New and Old World monkeys are so alike that few people seeing them side by side in a zoo can tell them apart, in spite of the established fact that their respective evolutionary courses have been independent for some forty million years. This process whereby two separate stocks within the same order occupying similar habitats have evolved along similar lines is known as *parallelism*.*

The terms parallelism and convergence are usually used in relation to structure, but there is no reason to doubt that behaviour is subject to parallelism in precisely the same way. Thus there is still hope that—genetic continuity apart—studies of living chimpanzee behaviour can throw some light, by analogy (through parallelism), on human behaviour, since the two families are so closely related.

There is overwhelming evidence of man's consanguinity with the two African species of great apes, the chimpanzee and gorilla, as anatomical studies and the biochemical analyses of certain body proteins have unequivocally shown. Anatomically, man and the African apes are fundamentally similar although superficially different. Ape locomotion is a highly special-

Ill. 48. A simplified classification of living primates. There are 55 genera of primates and 197 different species.

* *Convergence* is the term used to denote the same mechanism operating between different orders of mammals.

Ill. 49

60

ized affair and although neither living chimpanzees nor living gorillas are habitual brachiators, arboreal arm-swinging has in the past played a major role in shaping those physical proportions in which they differ so strikingly from man. Behaviourally, the situation is somewhat less clear. There are many aspects of chimpanzee communication—facial expression, manual gestures, postural activities, greeting and reassurance gestures—which have obvious analogies with their human equivalents and which should be regarded as behavioural parallelisms. Even the most disciplined scientist, for whom the idea of anthropomorphism is barely less obnoxious than devil worship, cannot fail in his off-duty moments to be emotionally impressed by the human qualities of the day-to-day behaviour of a group of chimpanzees. Then of course there is tool-using and tool-making. In the past, much has been made of the cultural significance of these activities; in fact for many years, not so long ago, evidence for tool-making was considered as evidence for humanity. The phrase 'man-the-toolmaker' was originally coined by that great American scientist and diplomatist, Benjamin Franklin. In more recent years, K. P. Oakley of the British Museum (Natural History) revived the concept, setting up tool-making as the principle criterion of humanity.[19] Now we know—principally through the studies of Jane van Lawick-Goodall and her students—that chimpanzees *use* and even, in a primitive way, *make* tools (see pp. 148, 150). Like so many supposed distinctions between man and his primate forebears, the positiveness of black or white turns on closer inspection into the neutrality of grey. This applies to hand function, to bipedalism, and, in fact, to all the criteria that have ever been proposed as evidence of behavioural discontinuity between apes and men.

There is one major exception to this general statement and that is the matter of language. The most intensive efforts have been made to teach chimpanzees to speak, all to no avail. Scientists now admit that, in spite of the possession of a speech apparatus that is virtually identical to man's, and a brain that, though smaller, differs in no known essentials from the human brain, there is a real discontinuity here between men and apes. The fact that apes have a wide range of calls, including barks, hoots, screams, soft pants and grunts, does not in any way invalidate the argument. The essence of

Ills. 136, 138, 141, 145–147
Ills. 137, 148

Ills. 143, 149, 150

Ills. 125, 152–154

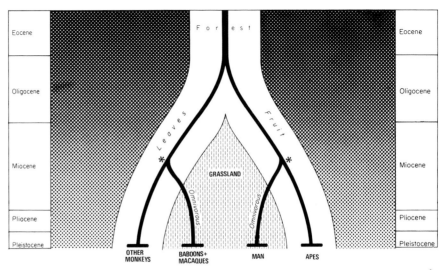

Ill. 49. A theoretical schema demonstrating how the descendants of the Old World primate lineages (monkeys and apes) converged in ecological terms, thus providing the basis for the evolution in baboons and man of certain analogous structural and behavioural characters. * indicates the common ancestors of these derived groups. 'Grassland' embraces both woodland and 'open' savannah.

speech and language does not lie simply in the factual or even emotional information that non-human animal calls can communicate, but in a number of unique factors of human speech that have been called by Charles Hockett 'design-features'. Essentially, such design-features allow the communication of abstract ideas, the discussion of events displaced in space and time, the transmission of information through teaching and learning, and the invention of new words. Human speech is not stereotyped like the closed call-systems of chimpanzees, but open and fluid.

Because hypotheses derived from analogy are almost as edifying as those derived from homology, the forward extrapolation strategy from primates to man can be extended to include species that are less closely related to us than the African apes.

Baboon societies, which are structured on the hierarchical system and *Ill. 50* achieve internal stability through adult male dominance behaviour, show many parallels with human societies. Early hominids in their formative years were probably open-country livers as baboons are today. Their situation with regard to group structure, food-getting and predator avoidance may well have been similar. Eventually, of course, man's peculiar talents as a weapon-toting, semi-carnivorous biped pushed him into a different ecological bracket but not before, one might surmise, certain open-country, baboon-like traits, such as male aggressiveness and dominance behaviour, had left *Ill. 51* their mark on his genetic constitution.

The strategy of forward extrapolation from non-human primates, though far from perfect, offers the best chance of success of all the options. Backwards extrapolation from 'primitive' populations of *Homo sapiens* can certainly help. Both types of study should receive the sort of fiscal encouragement that in our present society is reserved for half-baked sociological projects that are often more concerned with treating the symptoms than eradicating the disease.

The Hominidae are the zoological family which includes all fossil men and the one living species, *Homo sapiens*. The custom is to anglicize the term and refer to them as hominids (just as Pongidae, the great apes, become pongids in ordinary discussion). There have been many contenders for the title of the earliest hominids and some of these claims will be discussed below but, wishful thinking aside, the fact is that five million years ago is about the earliest date at which hominids can be positively identified. Certain fossils,

Ill. 50. Baboon societies are controlled by an oligarchy of adult males. Here, a high-ranking male baboon is being groomed by a female.

Ill. 51. An adult male baboon 'yawning', not from boredom but as a threat display in which the long, sharp canine teeth are revealed. This is analogous to an aggressive human being in the course of a heated argument taking off his jacket and rolling up his sleeves.

such as the genus *Ramapithecus* from the Dhok Pathan deposits of the Siwalik Hills of northern India and the closely related genus *Kenyapithecus* from East Africa, dating back to twelve and fourteen million years respectively, may represent early, prehuman, stages of the human family, but it is difficult to be certain about this without annectant or bridging forms to indicate the characteristics of early hominids.

In this chapter we shall be following the trail of man from his earliest beginnings to the full flowering of the genus *Homo*. The trail will take us to India, the East Indies, China, Europe and, above all, to South and East Africa.

Charles Darwin's intuitive genius is nowhere better exemplified than by his prophecy that the home of mankind would turn out to be the African

continent. When it is realized that in 1871, when Darwin's prophecy appeared   *Ill. 28*
in *The Descent of Man*, not one single human or near-human fossil had been
discovered outside Europe, the almost supernatural clarity of his vision is
staggering.

Today, Africa, particularly East Africa, is turning out human fossil
remains with the despatch of an assembly line, yet fifteen years ago the discovery of a single tooth with hominid characters was rare enough to merit a
letter to *Nature*. The richness of the East African scene today is due to the
genius and unbounded enthusiasm of one man, Louis Leakey. Without his   *Ill. 74*
astuteness and determination, our knowledge of the course of human
evolution would be infinitely poorer.

To the modern student, East African sites such as Olduvai, East Rudolf,
Omo, Kanapoi and Lothagam are magical names evoking the same en-
thusiasm that the South African sites of Taung, Sterkfontein, Makapan,
Swartkrans and Kromdraai engendered two decades ago.

The importance of *Australopithecus* discoveries in South Africa cannot be
overestimated. From 1924 onwards *Australopithecus* provided the evidence
that man actually had a past and not simply a history. Australopithecines are
not hominids of the genus *Homo* but proto-men of the genus *Australopithecus*. Like the apocryphal inventor of a non-alcoholic drink called 6-up,
they were within striking distance of success but never knew it. The South
African scene was originally the brain-child, once again, of one man,
Raymond Dart.   *Ill. 52*

Now Professor Emeritus at the University of Witwatersrand, Dart set the
anthropological world by its unwilling ears with the discovery in 1924 at
Taung, Bechuanaland, of a fossilized skull of a young australopithecine. For
twenty years, the Taung child was dismissed as a fossil ape by scientists still
bedazzled by a much more respectable ancestor for man discovered in 1909
by Charles Dawson at Piltdown. A fossil man from Sussex was more than a
match for a fossil ape from the colonies. Soon after the Second World War,
thanks principally to the scientific influence of Le Gros Clark, the tables
were turned. Piltdown man sank into shamefaced oblivion and the Taung
child and its notable contemporaries and successors rode tall, providing
genuine evidence for the existence of a prehuman stage of human evolution.

It may have been Charles Darwin who suggested that Africa was the
nursery of mankind, but it was Dart and Leakey who provided the evidence.

Although Africa, India and Europe were the principal incubators of the
placid egg that grew into an unruly offspring, the seeds of man were planted
some seventy million years ago when the first primates evolved from a
primitive mammalian stock of insectivore-like creatures. The zoological
Garden of Eden is as hard to pinpoint geographically as its biblical counterpart. Early fossil primates are found in North America and Western Europe,
including England; and since several genera are common to America and
Europe, the assumption is that the two continents were sufficiently connected at that time to allow a free exchange of fauna.

We can assume, as a working hypothesis based on the known distribution
of early primates, that they evolved in the Northern hemisphere—in Asia,
Europe and North America. In the language of continental drift theory, this   *Ill. 52*. Opposite: Professor Raymond
vast land mass is termed Laurasia. To narrow it down to Eurasia or to North   Dart, the discoverer of *Australo-*
America is more difficult. The odds are slightly in favour of North America   *pithecus*, holding the Taung specimen.

64

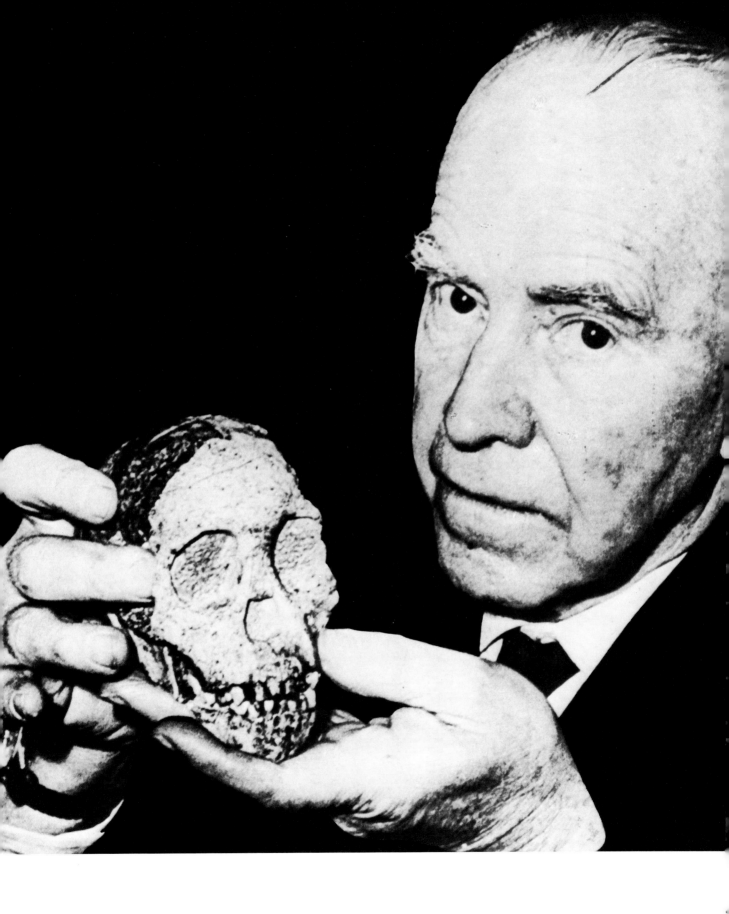

inasmuch as the earliest primates yet recognized of the species *Purgatorius ceratops* have been found in Montana. On the other hand, one of the earliest skulls of Eutherian mammals has been discovered in the early Cretaceous of the Gobi Desert of East-central Asia. *Zalambdelestes* is an insectivore-like creature, a putative forerunner of primates. Incidentally, this discovery of a placental mammal at the same geological time horizon as the fossilized eggs of the dinosaur *Protoceratops*, another discovery of the American Museum 1923 Expedition, was no doubt partly responsible for the dubious theory that the extinction of the dinosaurs at the close of the Mesozoic was hastened along by a sort of blitzkreig on the part of mammals to eradicate dinosaurs by eating their eggs.

Ill. 54

According to the evidence of continental drift theory, the northern reaches of the Atlantic between Greenland and Europe began to open up along an old fracture zone about fifty-five million years ago, thus severing the western bridge between Europe and North America. At about the same time, a northerly shift of Laurasia from an essentially subtropical location towards colder latitudes led to a cooling and a vegetational change in the eastern link between North America and Asia. The present Bering Strait and Aleutian Island chain mark the site of this ancient link. Free interchange of primate fauna via the Bering Strait area ceased for climatic reasons (primates are essentially tropical or subtropical animals) and, since early in the Eocene, North American primate stocks have evolved independently of stocks of Eurasian origin. The evidence for this zoological schism is apparent in the existence of two major subdivisions of the higher primates—the platyrrhine monkeys of Central and South America and the catarrhine monkeys of the rest of the world. There are two schools of thought as to the origin of the platyrrhines and, although this is somewhat of a diversion from the theme of this chapter, it is worth pursuing for a while. Our immediate relatives may be our prime concern but it would be uncharitable to neglect our more remote connections, however far they have strayed from the path.

Ill. 53

South America was an island continent during much of the Tertiary period. In terms of continental drift theory, South America can be visualized as breaking away in Jurassic times (200 million years ago) from Gondwanaland, the super-continent of the southern hemisphere. In early Tertiary times it is clear

Ill. 53. Continental drift. Left: during the Jurassic, 150 million years ago, the ancient world—Pangaea—broke apart, foreshadowing the continents of the present day. Right: the state of affairs in the Late Cretaceous, 90 million years ago. North America, Asia and Europe were still part of a single land mass—Laurasia—and India was moving away from Africa–Arabia towards an Asian landfall.

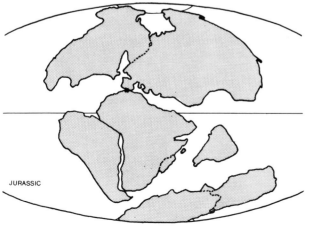

JURASSIC

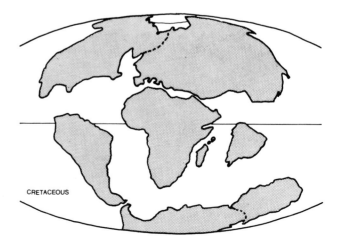

CRETACEOUS

			Duration in Million Years
CENOZOIC	Tertiary	Palaeocene	11
		Eocene	18
		Oligocene	13
		Miocene	18
		Pliocene	3
	Quaternary	Pleistocene	2
			65M

Ill. 54. The Cenozoic time-scale—the age of mammals.

from the fossil record that the ancient inhabitants of South America (some marsupials, sloths and extinct ungulates) had the continent to themselves. Between the late Eocene and early Pliocene times, three new groups of mammals reached South America—the caviomorph rodents, the platyrrhine primates and the Procyonidae (the raccoons and their like). This limited immigration suggests that their arrival owed much to chance. The chance factor in this case was so-called 'rafting', a passage facilitated by floating vegetation and favourable ocean currents. Floating islands of vegetation are not the desperate expedients of frustrated zoologists but real phenomena that have been observed time and time again.

An alternative theory must be given serious consideration. Studies of rock palaeomagnetism have made it possible to make a preliminary assessment of the date and rate of break-up of Gondwanaland. In the late Eocene, South America may have been as close to Africa as to North America. Thus the African origin of early platyrrhine monkeys is a possibility to bear in mind, particularly in view of the close similarity in structure between Old and New World monkeys.

As I have already said, the origin of New World monkeys, or platyrrhines, is really irrelevant in the present context but it is a pointer to the way in which we may have to revise our views on the origin and dispersal of primates as the facts of continental drift become more and more irrefutable.

Eurasian primates subsequently migrated in a southerly direction. In the beginning it is assumed that they reached Africa and South-east Asia; later they came to occupy peninsular India when that continental mass, initially part of Africa, had drifted north-eastwards to join up with Southern Asia in Miocene times. The colonization of the Southern Hemisphere was a necessary re-location of animals driven from Eurasia by the northward drift of the Laurasian super-continents. Primates today are tropical animals, as they were sixty to sixty-five million years ago. With the exception of a few mavericks, living primates occupy habitats within 30° north and south of the equator.

The quest for man starts with the earliest primates. As a genus, man, or something like him, has always been on the cards. A higher form of primate life is the logical culmination of the evolutionary trends that characterize the order as a whole. As it happens, a higher form of higher primate (apes and monkeys are often referred to as higher primates to distinguish them from the lower primates, the prosimians) has developed twice to our knowledge. First came the apes—distinctly emancipated monkeys—and, then, out of the apes arose man.

In view of these considerations, a brief summary of primate evolution *before* the coming of man and his hominid ancestors would be of considerable value if only to identify those components of the human make-up that constitute the legacies of his humbler past.

It is fundamental to the understanding of the nature of man to appreciate that his heritage, in ecological terms, is a dual one. Modern man is derived from a ground-adapted stock and we have reason to suppose that his immediate forebears have been ground livers for several million years. But his remote ancestors, the unknown monkey-like or ape-like primates which preceded him, were totally arboreal.

There are many species of primates today for whom the ecological designation 'arboreal' or 'terrestrial' ill reflects the pattern of their lives, which embraces both zones. *Ill. 58* demonstrates the habitat distribution of some living monkey and ape genera of Africa; it can be seen that there is a spectrum of habitat utilization ranging from the wholly arboreal primates, which never set foot upon the ground, to the wholly terrestrial, which never climb a tree. Man, surprisingly enough, is not the most terrestrial of primates. For one reason or another, he is often to be found above the ground in trees, on ladders, on scaffolding, or even up at the top of telegraph poles. The gelada baboon on the other hand, which lives at high altitudes in Ethiopia and in terrains that have no trees at all, is totally grounded. Even 'terrestrial' primates like the common baboon sleep in trees at night in most areas in which they are found and utilize them as a food source during certain seasons when figs and fever trees are in fruit. As might be expected, such habitat differences are reflected in morphological and behavioural differences.

Prior to the emergence of the hominids, an event for which the *primum mobile* was the adoption of ground living, the primate habitat was exclusively arboreal. Many of the physical characteristics that denote a primate and distinguish him from other groups of mammals are associated with arboreality; these include, for example, the opposable thumb, the forward-facing eyes, a large brain and a bodily structure adapted for the upright posture. Other groups possess some of these characters but only primates possess them all.

By becoming ground living, the hominids did not acquire any *new* characters, they merely improved on the old ones in the course of evolution. As structure became more complex, so functional capabilities became extended and the behavioural repertoire more elaborate. Hominids grew better thumbs, larger brains, and more refined postural and walking mechanisms.

Ills. 57, 58

Ills. 55, 56

Ills. 55, 56. Both baboons and man are essentially ground livers, but neither is wholly terrestrial. Baboons sleep and often feed in trees when certain fruits are in season (below left), while man's arboreal heritage makes the climbing of trees both an attractive and facile activity, particularly when it enables him to watch a football match free of charge!

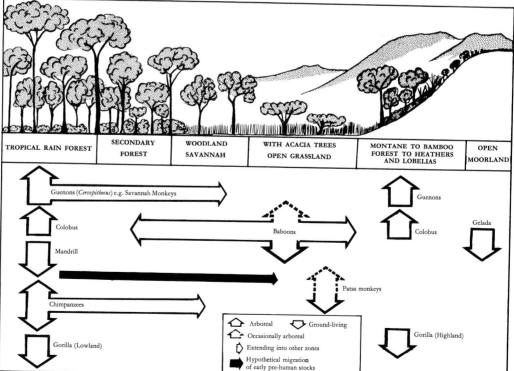

TROPICAL RAIN FOREST	SECONDARY FOREST	WOODLAND SAVANNAH	WITH ACACIA TREES OPEN GRASSLAND	MONTANE TO BAMBOO FOREST TO HEATHERS AND LOBELIAS	OPEN MOORLAND

Guenons (*Cercopithecus*) e.g. Savannah Monkeys

Guenons

Colobus

Baboons

Colobus

Gelada

Mandrill

Patas monkeys

Chimpanzees

Gorilla (Lowland)

Gorilla (Highland)

Arboreal   Ground-living
Occasionally arboreal
Extending into other zones
Hypothetical migration of early pre-human stocks

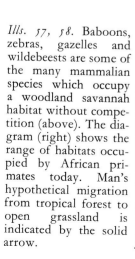

Ills. 57, 58. Baboons, zebras, gazelles and wildebeests are some of the many mammalian species which occupy a woodland savannah habitat without competition (above). The diagram (right) shows the range of habitats occupied by African primates today. Man's hypothetical migration from tropical forest to open grassland is indicated by the solid arrow.

69

On the face of it, it seems paradoxical that man's distant arboreal heritage should have prepared the way so conveniently for his career as a ground liver. But there is no real paradox; the exigencies of life in the trees 100 ft. above the ground demand the highest degree of manual, visual, locomotory and mental precision, just as they do in the life of man, who faces dangers not just simply of falling (although such accidents are common enough), but in the factory, in the home, or when driving a car, to say nothing of the numerous sociological and psychological hazards which can affect his well-being.

EVOLUTION OF THE HOMINOIDS

The earliest primate fossil is said to be that of *Purgatorius* from Purgatory Hill, Montana. To the non-palaeontologist this designation might seem to be more of an act of faith than a legitimate scientific judgement, since it is based on a single upper molar tooth. Even to a hardened palaeontologist such a bold diagnosis seems to be a bit of a presumption, since the likelihood of similarity being due to convergence must be very high indeed.

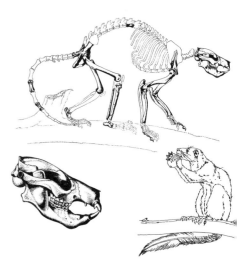

Palaeocene and Eocene. The fossil family Plesiadapidae is well known from a number of specimens of late Palaeocene and early Eocene date found in North America and Europe. Some of the Plesiadapidae, *Plesiadapis* itself for instance, were quite large, almost cat-sized. Once again, the primate affinities of *Plesiadapis* are based on dental similarities of the molar teeth with an undoubted Eocene primate called *Pelycodus*. The cusps of the molars are low and rounded, indicating that they have a herbivorous-frugivorous diet rather than a purely insectivorous one.

Below the neck, *Plesiadapis* is not at all primate-like. The relative length of the fore- and hindlimbs is more typical of carnivores or rodents, or any other ground liver. The hands and feet carry claws instead of the flattish nails of typical primates. Some authorities[4] have recommended that the Plesiadapidae be excluded from the primates on the grounds of these differences.

However, I believe it would be wrong to exclude *Plesiadapis*, or for that matter *Purgatorius*, from the order for these reasons alone. Dr. Szalay of the American Museum of Natural History would go further and say that the teeth of *Plesiadapis* display a functional shift away from a basic mammalian insectivorous diet to a herbivorous-frugivorous one typical of arboreal primates.

Such a dietary shift is likely to have accompanied a behavioural switch from a nocturnal to a diurnal way of life. Most, though not all, primates are awake during the day. The nocturnal primates are, with a single exception, lower primates and generally speaking they are small in size. For the most part they are extremely active, for which an insect diet, consisting of highly nutritive animal proteins, is essential.

The earliest primates to show arboreal adaptations and a generally modern aspect were the Adapidae and the Anaptomorphidae. The adapids were a large family of lemur-like prosimians widely distributed in North America and Europe. The earliest is called *Pelycodus* and is known from the delightfully named Lostcabinean period of the early Eocene. *Pelycodus* shows some dental affinities with both the earlier *Plesiadapis* and the later *Notharctus*. The North American adapids *Notharctus* and *Smilodectes* were clearly well adapted for an

Ills. 59, 60. Reconstructions of the Palaeocene primate *Plesiadapis* (above), and of the Eocene primate *Smilodectes*, a close relative of *Notharctus* (below).

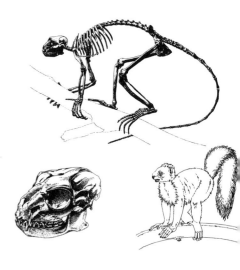

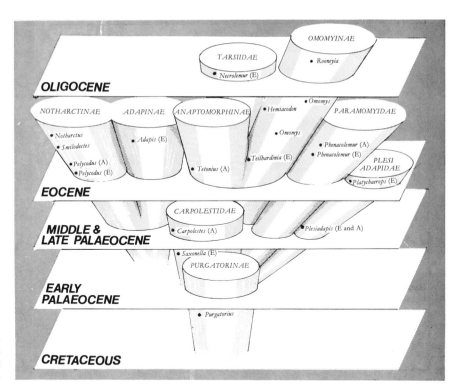

OLIGOCENE

OMOMYINAE
• *Rooneyia*

TARSIIDAE
• *Necrolemur* (E)

NOTHARCTINAE
• *Notharctus*
• *Smilodectes*
• *Pelycodus* (A)
• *Pelycodus* (E)

ADAPINAE
• *Adapis* (E)

ANAPTOMORPHINAE
• *Tetonius* (A)

• *Hemiacodon*
• *Omomys*
• *Omomys*
• *Teilhardinia* (E)

PARAMOMYIDAE
• *Phenacolemur* (A)
• *Phenacolemur* (E)

PLESIADAPIDAE
• *Platychaerops* (E)

EOCENE

MIDDLE & LATE PALAEOCENE

CARPOLESTIDAE
• *Carpolestes* (A)

• *Plesiadapis* (E and A)

• *Saxonella* (E)

PURGATORINAE

EARLY PALAEOCENE

• *Purgatorius*

CRETACEOUS

Ill. 61. Diagram to show the principal groups and some of the known genera of Palaeocene and Eocene primates. A = America, E = Europe.

arboreal life, having hands and feet that were well suited to grasping: the ends of the fingers and toes were surmounted by nails rather than claws. The presence of nails in itself is not a unique primate characteristic. Although almost all primates possess them, so do a number of non-primate creatures, such as the tree hyrax and some tree-climbing marsupials.

At one time W. K. Gregory, the great American zoologist, believed that *Notharctus* might have been ancestral to the South American or New World monkeys, but this view was based on a misconception and receives little or no support today. The American adapids became extinct towards the end of the Eocene. The European adapids, on the other hand, persisted longer and theoretically could have provided the stock from which living lemurs and lorises evolved.

There is only one family of Eocene prosimians that has been proposed as possible ancestors for monkeys and apes, but the essential evidence for this is simply not there. The family in question are the Anaptomorphidae, which contains three subfamilies. One of these, the Omomyinae, were rather short-faced primates with a somewhat monkey-like dental apparatus. Apart from foot bones from one genus (*Hemiacodon*), which indicate a springing or leaping type of locomotion similar to that of living galagos and tarsiers, there is little to be learnt from the fossil record about the way of life of the omomyines. It has been tacitly accepted that the European omomyines were the most likely candidates for the unknown ancestors of Old World monkeys. Recently this possibility has been seriously questioned and a new candidate called *Pelycodus* has appeared. *Pelycodus*, of the earliest Eocene, occurring in both European and North American deposits, has structural links with *Plesiadapis* on one hand and with the early catarrhine primates such as

71

Oligopithecus (see below) on the other. Thus we see the first glimmerings of a possible lineage: *Plesiadapis–Pelycodus–Oligopithecus*, but it is a fitful light much candlepower removed from the brilliant illumination of established fact.

So we move into the next phase of primate evolution without any very clear idea of where we have come from. All we know is that our Eocene ancestors were agile arboreal forms, mainly herbivorous and frugivorous in diet with prehensile hands, well-developed eyesight and a progressive type of brain.

The scene changes. Up until now the locale has been Holarctica, the northern hemisphere. As these regions drifted northwards they became less attractive for primates to inhabit. Fruit and leaves became seasonal and the day-length progressively shorter in the winter—conditions that impose severe restrictions on the hours available for finding food.

Since the beginning of the Palaeocene, temperatures in North America and Eurasia had been steadily dropping. Much of the information relating to ancient climates can be gleaned from the study of fossil plants, the 'thermometers of the ages' as they have been called. During the Eocene, mean annual temperature in Seattle, Paris and London were comparable to those experienced in Mexico City today. Subtropical forests reached as far as 53°N. There was a sharp drop in temperatures during the late Oligocene and throughout Holarctica temperate forests were replacing tropical forests, and mean annual temperatures were in the region of 64°F—conditions that one might today experience in the southern states of the U.S.A.

Ill. 62

During the Miocene the downward trend of temperatures in the Northern Hemisphere continued. In the Ukraine, for instance, deciduous trees of temperate forest type were succeeded by coniferous-deciduous woodlands. The changing geomorphology of the Miocene, the epoch of mountain-building, resulted in major changes in regional climates during which forests became grasslands, and grasslands became deserts.

The zones of primate occupation shifted southwards towards the equator as the northern latitudes became less desirable and, from this time on, the principal fossil localities are to be found in Africa, South-east Asia and the more southern regions of Eurasia such as India, Italy and Greece. The Fayum region of Egypt, now arid desert, marks the site in the Oligocene of tropical forest well supplied with watercourses and deltaic swamps. Our

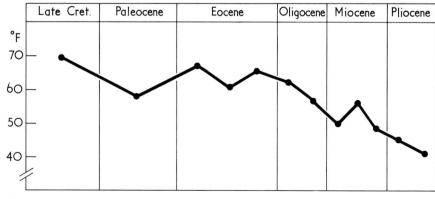

Ill. 62. Mean annual temperature changes at latitude 40°–50°N throughout the Tertiary.

Ill. 63. Collection of fossil primates in the Fayum, Egypt, by the Yale expedition of 1962. Dr. Elwyn Simons (centre) at the site of the discovery of the oldest known fossil ape, *Oligopithecus savagei*.

knowledge of the earliest Old World primate fauna comes almost entirely from the Fayum, which has been the site of various fossil expeditions since the initial visit by the American Museum of Natural History in 1906–7. The most recent work, carried out in the early 1960s with great success by a series of Yale expeditions, was headed by Professor Elwyn Simons.[23]

Ill. 63

The variety of genera and species of primates that have been identified in the Fayum is really quite remarkable. Among the discoveries, which incidentally could all be placed in a shoe box with room to spare so small are the specimens, are the possible ancestors of monkeys, gibbons, great apes and man. Most of the specimens consist of jaws, teeth and fragments of the skull. Of the many hundreds of separate items only one more-or-less complete skull has been discovered. Limb bones are scarce, as indeed they are at most fossil sites; the bones of the legs and arms are rich in calcium and marrow and are delectable food sources for leopards, vultures, porcupines and hyenas. There are vast accumulations of limb-bone slivers and fragments in the caves at Swartkrans in South Africa, for example, which in the view of Dr. C. K. Brain, the Director of the Transvaal Museum, were the leftovers of leopards' lunches (see pp. 95, 96).

The absence of limb bones at the Fayum sites may have another explanation. At the time when the fossil beds were laid down, the Fayum was an area of tropical rain forest criss-crossed by numerous watercourses. Dead animals on the forest floor are subject to attack by ants, termites and beetles which dispose of the flesh, and to the acid nature of the soil which decalcifies the bones. Bearing all these hazards in mind, it is a wonder, really, that palaeontologists have anything to work on at all. Some of the best fossils are of animals that have been preserved by accident. Animals that have fallen into tarpits, or rivers or lakes, been buried by volcanic ash or been washed into deep rocky crevices and caves.

There is a saying amongst palaeontologists that God made the skull and the Devil made the teeth. This is simply a way of indicating that teeth are very complex structures and that their interpretation is a permanent source of disagreement and dispute. This is unfortunate, as teeth are by far the most common items to be found at fossil sites; being unsavoury and indigestible food for predators, they have a way of surviving when all other skeletal parts have disappeared.

In addition, they are durable in another sense inasmuch as during evolution they tend to change their form and function very slowly indeed. Thus, they provide an invaluable source of information for determining phylogenetic relationships between fossil species.

Oligocene Hominoids. Four hominoid genera have been identified in the Fayum: *Oligopithecus*, *Propliopithecus*, *Aeolopithecus* and *Aegyptopithecus*, which have to be considered as being on or near the ancestry of modern apes and therefore of man.

Ills. 64–66

Propliopithecus has been a somewhat enigmatic genus since the discovery of a single lower jaw by Richard Markgraf in 1908. The specimen was studied by the German palaeontologist Schlosser in 1911 who pronounced it to be an ancestral gibbon, but at the same time he started a hare, which has been running in stops and starts ever since, when he tentatively suggested that alternatively *Propliopithecus* might be near the ancestry of man. Ten years ago, Dr. Elwyn Simons of Yale expressed a similar belief but has currently withdrawn from this position, as have a number of his colleagues. If *Propliopithecus* is not a hominid ancestor perhaps it is a hominoid ancestor. There does not seem to be any evidence of special relationship to the gibbons so perhaps it is best left in limbo, with the vague generalization that it may have been on or near the root of the newly evolving ape-cum-human superfamily, the Hominoidea.

Aeolopithecus, once again, is represented by a single lower jaw complete except for the incisor teeth. The official view is that *Aeolopithecus* is an ancestral gibbon for reasons that are not particularly convincing. What is certain, however, is that it was a small ape, arboreal and fruit-eating, so there is no denying that it might ultimately have turned into a gibbon, which is by far the smallest of modern apes and is certainly both a fruit-eater and an arborealist.

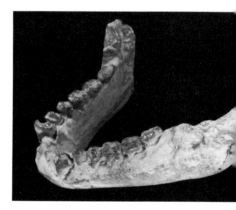

We are on stronger ground with the fossil genus called *Aegyptopithecus*. Not only are there a number of specimens of both jaws available but there is an almost complete skull discovered in 1966. In the form of its teeth, *Aegyptopithecus* is typically ape-like in nearly all respects and shows striking similarities to *Proconsul africanus*, an exceptionally well-known species from the Miocene of Africa whose antecedency to modern apes is hardly in doubt.

Oligopithecus appears to have been a very primitive ape. Its dental formula, that is to say the number of each type of tooth in each half of each jaw, can be expressed thus: *Incisors* $\frac{2}{2}$; *Canines* $\frac{1}{1}$; *Premolars* $\frac{2}{2}$; *Molars* $\frac{3}{3}$ or, more simply: $\frac{2 \cdot 1 \cdot 2 \cdot 3}{2 \cdot 1 \cdot 2 \cdot 3} \times 2$. The factor of 2 gives one of the total number of teeth, which

Ills. 64, 65. Lower jaws of two Oligocene primates. Above: *Propliopithecus*, discovered by Markgraf in 1908. Below: *Aeolopithecus*, from Upper Fossil Wood Zone, Yale Quarry I, Fayum, Egypt. *Aeolopithecus* is regarded as an ancestral gibbon.

is 32. This is the dental formula of all Old World primates—monkeys, apes and man. New World monkeys and prosimians have 36 teeth, thus their dental formula is $\frac{2 \cdot 1 \cdot 3 \cdot 3}{2 \cdot 1 \cdot 3 \cdot 3} \times 2$. This formula is regarded as more primitive. There are two genera found at the Fayum that still retain the primitive 36-tooth formula and these are thought to be monkeys and not apes. The fossil apes like *Aeolopithecus* and *Aegyptopithecus*, on the other hand, possess the 32-tooth formula. The fact that *Oligopithecus* has the 32-tooth formula is one of the reasons that it is placed with the apes and not with the monkeys.

Ill. 66. Skull of *Aegyptopithecus zeuxis* from Yale Quarry M, Fayum, Egypt.

Thus, *Aegyptopithecus* must be regarded as the earliest (twenty-eight million years old) clue in our quest. If it is ancestral to the apes it must assuredly be ancestral to man as well. We left the Eocene with a possible lineage in mind that led from *Plesiadapis* through *Pelycodus* to *Oligopithecus*. Can we pursue the lineage from *Oligopithecus* through to *Aegyptopithecus*? Not really, unfortunately; but neither can we disprove the possibility.

Miocene Hominoids. In this chapter I have tried to present primate evolution in the context of the environment and to point out how the global changes in climate, vegetation and geology are the direct and indirect consequences of continental drift; and how such changes have affected the distribution of primates, the nature of their habitats, the patterns of their behaviour, and the course of their evolution. At no time were these factors more relevant than during the Miocene epoch—relevant not only for primate evolution as a whole, but for human evolution in particular.

During Miocene times volcanic activity, rift-valley formation and mountain building, the outward expressions of continental movements, were in full swing. One of the effects of these events was the spread of grasslands at the expense of forest cover. Grasslands—or savannahs, llanos or prairies as they are called regionally—offered new evolutionary opportunities to a variety of animals which previously were adapted to forests and woodlands. As is well known, the horses took full advantage of this opportunity and evolved into large, single-hooved, fast-running, grazing quadrupeds. But more importantly for us, the primates, for whom opportunism has been the keynote of their success, also took full advantage of the opening of new horizons.

There are two groups of Old World or catarrhine primates which have adopted a ground-living way of life—the baboon-macaque group and the hominids. Other major groups of primates, including the langurs, the most arboreal of Old World monkeys, have given rise to the odd ground-living form. So have the lemurs; the ring-tailed lemur is an example of a facultative ground liver. The only major group that cannot count a ground liver amongst its ranks are the platyrrhines or New World monkeys. Why this should be the case is a zoological puzzle. The most likely explanation is zoogeographical rather than zoological as the following excerpt indicates:[17]

> Old World primates in the Middle Tertiary [Oligocene and Miocene] occupied a vast range of territory in Eurasia which is now warm-temperate or temperate, but which in the Miocene was subtropical. The cooling of late Cenozoic times had its effect primarily on areas outside the present tropics. From these considerations the inference might be made that the [ancestors] of living Old World primates first evolved in parts of Europe and Asia well removed from the Equator and thus were the first to seize the opportunity of a novel ecological niche presented by the emerging grasslands. Consider now the geographical range of New World monkeys, which from late Eocene times had been restricted to the Latin-American continent well within the climatic zones that today we regard as tropical or subtropical. It is generally agreed that there was little, if any, change in climatic conditions during the Miocene or Pliocene that could have affected the distribution of South American forests. Consequently no

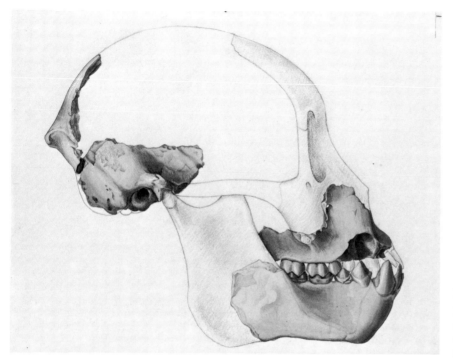

Ill. 67. Proconsul africanus. This reconstruction by the author and Professor P. R. Davis is based upon fragments of the skull derived from two specimens. Only the unshaded portions are hypothetical.

new ecological opportunities, comparable to those offered to Old World primates, presented themselves to the platyrrhines in mid-Tertiary times. Thus they have remained an essentially arboreal stock.

The baboons and the macaques are closely related animals which share, amongst other things, a broadly common pattern of social behaviour. Historically, it seems probable that the common ancestor of these two geographically separated genera (*Papio* and *Macaca*) was a macaque-like creature which emerged during the Miocene about the same time as the distant ancestors of man—and under the same conditions.

Among the man-ape group, only man has adopted ground living as an habitual way of life, although it is true that both gorillas and chimpanzees are less arboreal today than they once were (male gorillas, surprisingly, never climb trees). However, they have never deserted the forest habitat and, thus, have been denied the selection effects of life in open country. The significance of the adoption of ground living for the evolution of man is discussed in Chapter 4.

In *The Times* of 30 October 1948 there appeared an announcement of 'outstanding importance'. Dr. L. S. B. Leakey reported that his wife Mary had discovered the skull of 'a Miocene ape of the species *Proconsul africanus* . . . that is estimated to be at least twenty million years old.' Shakespeare's description of Cleopatra—'age cannot wither her nor custom stale her infinite variety'—is superbly applicable to the 1948 skull. It is still of 'outstanding importance' and twenty-five years later has neither been deposed nor assassinated, which is more than can be said for a lot of major discoveries in this field.

Proconsul africanus from Kenya, East Africa, may still quite reasonably be regarded as the common ancestor of apes and man. If pressed, I and others

Ills. 68–70. Opposite: three stages in the reconstruction of the hand of *Proconsul africanus.* Left: as found. Bones have been partially 'developed' from the matrix. Centre: bones assembled. Right: skeletal hand reconstructed. The missing bones are unshaded.

76

who have studied and written about the Leakeys' skull, about their subsequent discovery of a limb skeleton of the same species, and about the earlier fragments of jaws and teeth, would undoubtedly agree that *Proconsul africanus* was probably not *the* common ancestor (being already more commited to an ape-future than to a human one) but as near to it as makes no essential difference.

To our tutored eyes *Proconsul* was an ape, but to its happily ignorant contemporaries it must have looked very like a monkey, with its quadrupedal gait and its long tail. There is no fossil evidence in fact that it had a tail but, in the light of the other parts of the skeleton that have been discovered, the existence of a tail is more likely than not.

Ill. 67
Ills. 68–70 Anatomically *Proconsul* was still rather more monkey-like than ape-like. In the form of its skull, the shape and size of the brain, its inferred locomotion and hand structure, it finds its nearest living counterpart in some of the less specialized langurs of India and South-east Asia and in the spider monkeys of South America, rather than with living apes. The dental structure of *Proconsul* is fundamentally of the ape type and differs from it only in the smallness of the upper incisor teeth, which lack the shovel-like shape of living pongids. Furthermore the molar cusp pattern is somewhat more primitive than in living apes. As many authorities have observed, the teeth of the Oligocene *Aegyptopithecus* are very similar to those of *Proconsul africanus*. As a working hypothesis, it is quite acceptable to look upon *Proconsul* as a descendant of this Oligocene species.

In the jargon of the last decade *Proconsul* is somewhat of a 'crazy mixed-up kid', at least to our eyes it is, having the teeth of an ape and the bodily form of a monkey; but in spite of being zoological chimera, *Proconsul* was clearly superbly well-fitted for the ecological situation of its time. Consensus

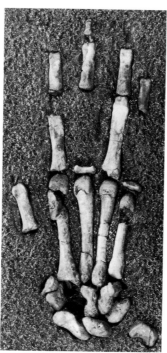
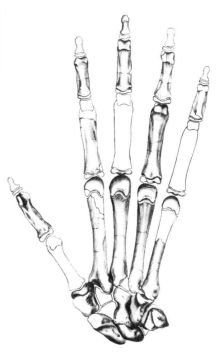

opinion is that *Proconsul africanus* was a Miocene ancestor of the living African apes.

There are two other *Proconsul* species, *nyanzae* and *major,* both less well known than *africanus. Proconsul major* was a much bigger creature and so gorilla-like in the parts of its anatomy that are known that some authorities even go so far as to say that it *was* a gorilla. This could only be acceptable if the term gorilla defined a way of life and not so much a species. Studies of blood proteins of living gorillas and chimpanzees indicate that it is only in the last few million years that they have evolved into separate forms.

Chronologically, the next important group of fossils are referred to collectively as the dryopithecines. These 'woodland apes', so-called, were widespread in time and space. The subfamily Dryopithecinae, as they are classified technically, are recognized from the middle Miocene to the lower Pliocene—a span of some eight to ten million years. Geographically their domain was Eurasia, extending from Spain in the west to India in the east. Clearly they were a highly successful group living in temperate and subtropical woodlands and forests. In fact, so many dryopithecine remains have been discovered—each being given a different latin name—that a study of the subfamily as a whole was a Herculean task of daunting proportions. Two scientists from Yale, Drs. Simons and Pilbeam, were not dismayed and nobly tackled the problem, with the result that of the extant twenty-five generic names only three were regarded as valid by the time the job was complete.[24]

The really interesting thing to emerge from Simons' and Pilbeam's revision was that out of this welter of woodland apes three clear-cut phylogenetic lines were apparent: (1) *Dryopithecus*: ape-like creatures with large canine teeth and heavy jaws. Probably arboreal in habitat and fruit-eating in diet. (2) *Ramapithecus*: man-like creatures with small jaws and moderately sized canine teeth. Probably terrestrial in habitat and vegetarian in diet. (3) *Gigantopithecus*: neither man-like nor ape-like but constituting a separate ecological type. Large in build with disproportionately immense jaws adapted to cope with a specialized diet of grasses and seeds.

Dryopithecus is regarded as the stock from which modern apes arose. *Ramapithecus,* which will be discussed more fully in Chapter 4, is a possible forerunner of man, and *Gigantopithecus,* ancestral to the giant 'apes' of China of the middle Pleistocene, is a much maligned form which is popularly supposed to be the antecedent of the Abominable Snowman of the Himalayas and of the Sasquatch of British Columbia and North-western United States. An interesting genus you must agree!

There are of course other primate fossils of the Miocene interesting in their own right, such as *Pliopithecus* of Southern Europe and *Limnopithecus* of East Africa which are credited, probably incorrectly, with the antecedency of the gibbons, and *Oreopithecus* from Monte Bamboli in Italy, a fascinating maverick which is neither ape nor man—nor Abominable Snowman. The monkeys appear in the fossil record towards the end of the Miocene with *Mesopithecus,* a macaque-like form from which living macaques, baboons and mangabeys may have evolved.

Some authorities, notably Simons and Pilbeam, classify *Proconsul* as a dryopithecine. I am not prepared to go as far as this, but nevertheless I readily admit that the *Proconsul* group provide the ancestral stock out of which later dryopithecines—and therefore living apes and man—evolved.

4 The Talented Primate

John Napier

THE stage is now set for the magnificent drama of human evolution and the time has come for the curtain to rise, the actors to be introduced and the play to unfold. Promising as this analogy is, unfortunately it can be pursued no further for, to tell the truth, the scenario has still to be written.

To build a coherent story of human evolution out of the scraps of evidence at our disposal is something of an act of faith. Not that there is any reason to doubt that evolution took place; it happened all right, but the way it happened, the *mise en scène*, the timing of the events and the identities of the principal characters are not known with any certainty. Perhaps this is over-stating the case but better this than to give the impression that all is known.

Most surveys of human phylogeny start at the beginning and work forwards. In this chapter, I propose to reverse the procedure and start at the end; the end at least is indisputable.

HOMO SAPIENS

Our own species has been in existence for about 250,000 years. The species includes four subspecies: *Homo sapiens sapiens* and *Homo sapiens neanderthalensis, Homo sapiens soloensis, Homo sapiens rhodesiensis*. Of these only *Homo sapiens sapiens* exists today. In the cause of simplification, the trinomial *Homo sapiens sapiens* is here reduced to the binomial *Homo sapiens* or, more simply still, to the colloquialism 'modern man'.

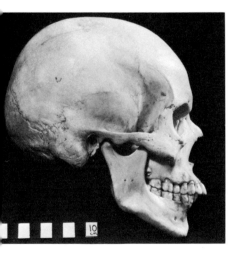

Ill. 71. The skull of modern man. Note the absence of brow ridges, the presence of a chin, and a high, curved skull vault.

Modern man, used in the Linnean sense, refers to the immediately post-last-glaciation populations, usually called the Cro-Magnons, and their descendants. However, some authors use the term to include all the subspecies listed above. Modern man is defined by the characters of the skull: a high vault with flattish sides, the greatest width being sited high on the sides between the two parietal bones; a brow that only exceptionally forms pronounced ridges over the eyes; a flat face which is pulled back so that it lies under the brain case. There are well-developed hollows on either side of the nose, and a chin is present. The forehead rises vertically from above the eyes.

There are, of course, a number of geographical varieties of modern man that are reflected in the skull features, but none of these are as great as the differences between modern man and Neanderthal man for instance.

Neanderthal Man. Neanderthal man is everybody's idea of a human ancestor. He is the original caveman with animal skins covering his squat, muscular body, and a brutal face with much the same cast of degeneracy as that shown in some of William Blake's profiles of Roman emperors. For good measure he is credited with uncontrolled possessiveness which led him into battles to the death with dinosaurs (extinct sixty-five million years before his time) in defence of his nubile and over-exposed mate. Cartoonists and film-makers have a lot to answer for.

Neanderthal man is exceptionally well known to anthropologists from numerous skulls and skeletons. Anatomically he was a good *sapiens* type differing only in matters of degree from modern man. The chief differences relate to the forward projection of the face, particularly the nose and upper jaw, the falling-back of the chin, and the lowness of the brow combined with a backward projection of the rear of the brain case, giving the classic 'bun-shaped' occiput. Neanderthal man owed many of his anatomical oddities to the rigours of the last Ice Age. They were a very specialized race of European 'eskimos' with short, sturdy bodies adapted for conserving heat, large knobbly joints, and massive jaws that could chew tough, partially-cooked food. Their front teeth were hefty, which has led some authorities[2] to argue that the incisors were in constant use as a sort of vice in the working of wood and skin; their molars were larger than modern man's but not nearly as massive as those of *Homo erectus*.

Although culturally Neanderthal man's activities were more prosaic than his successors, whose cave-paintings are legendary, he was well advanced and his characters were entirely suitable for the rigours of the Würm glaciation through which he lived. His 'tool-kit' was practical although lacking the refinements of his more leisured inheritors who invented the splendid concept of spare-time hobbies. Neanderthal man showed the beginnings of spiritual awareness: he buried his dead. Not only that, he buried them ceremoniously.

The popular image is way off target but no real harm is done thereby. More misleading is the belief that Neanderthal man was *ancestral* to modern man. As we shall see, this is quite incorrect.

In 1848, in Gibraltar, a *sapiens*-like skull was discovered in a quarry on the north face of the Rock. It went unrecognized for almost two decades, until the remains of a human creature found buried in the floor of a cave in the bluffs bordering the River Neander in Germany in 1857 led, in 1864, to the recognition of a new species of man, *Homo sapiens neanderthalensis*. Even then, another fifty years were to pass, during which many new discoveries were made, before the Gibraltar skull was accepted into the 'family'.

The climate of scientific opinion in 1864 (five years after the publication of Darwin's *The Origin of Species*) was a bit one-sided; it was Darwin and Huxley versus all the rest. Naturally, a well-publicized discovery of an ancient fossil man was received with distaste and, worse, with embarrassment that such an ancient human being should have such a large brain. The man of the Neander Valley was written off as a mental defective and probably criminal at that; or, with devilish ingenuity, as one of Napoleon's soldiers with water on the brain left dead by the roadside on the retreat from Moscow!

Unfortunately for the denigrators, Napoleonic soldiers with water on the brain subsequently cropped up in France at La Chapelle-aux-Saints in 1908 and in La Ferrassie in 1909, in Belgium, in Germany, in Rumania, in the Crimea, in Central Asia, in Palestine, in East, North and South Africa. The total number of known individuals of this race is now more than a hundred. Even in the more enlightened times of the early twentieth century, the scientific image of Neanderthal man was of a stooped, bent-kneed half-ape.

80

Ill. 73. Opposite: a Pleistocene summit. From left to right, Olduvai man, Java man, Neanderthal man, Cro-Magnon man and modern man meet to discuss the problems facing the human race.

Ills. 6, 38, 104

Ill. 72. Reconstruction of the head of Neanderthal man (cf. *Ill. 47*).

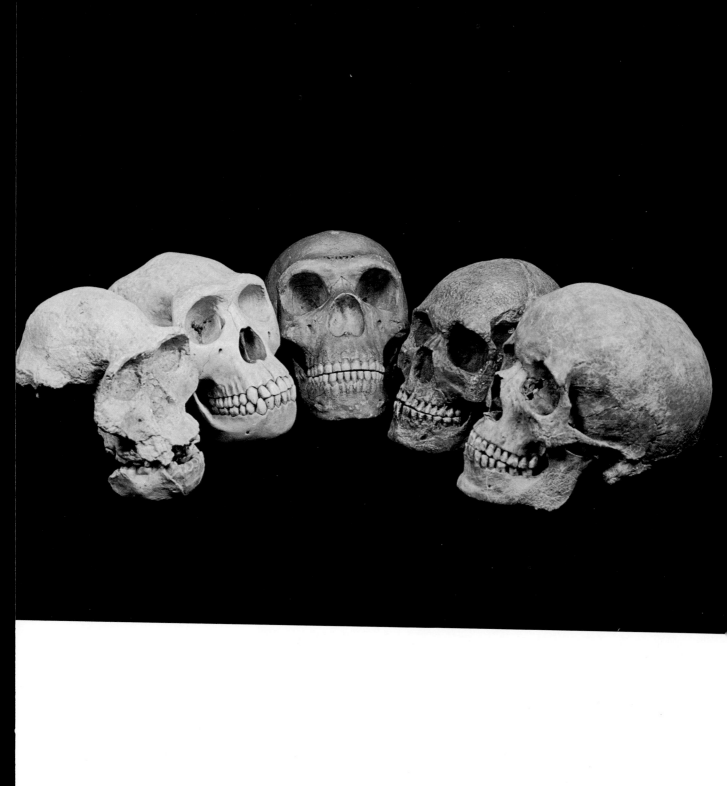

Ill. 74. Opposite: Louis Leakey at work on the skull of 'Zinj'.

Nasty, brutish, and short you might say. But that is past history and we now accept the men of the last Ice Age as being good members of our own species who put up a struggle for survival in appallingly adverse climatic conditions and lost. The 'Neanderthal questions' are now: where did they come from? How are they related to *Homo sapiens sapiens*? What happened to them? Most scientists seem to agree that while Neanderthalers were battling it out with cave bears and woolly mammoths in the cold environments of the last Ice Age, the predecessors of modern man were lotus-eating elsewhere. In Africa? Asia? the Middle East? We simply do not know.

Taking the last question first. When the glaciers of the Würm Ice Age receded, the cold-adapted Neanderthalers would have been at a disadvantage in competition with the more generalized Cro-Magnon people, and thus were liable to become extinct as a race. It seems that this is what happened. Two indisputable facts appear to support this point of view. Firstly, no Neanderthal fossils have been found younger than 30,000 years. Secondly, there is no cultural connection between the Neanderthals and the Cro-Magnons. For example, the 'tool-kit' of the Cro-Magnons was much more sophisticated, and in cave deposits sterile layers can be shown to exist between the remains of the two cultures, suggesting an absence of contact or continuity.

The extinction of a population may result from either war or economic attrition, or both. In neither instance is it likely that genocide would be total.

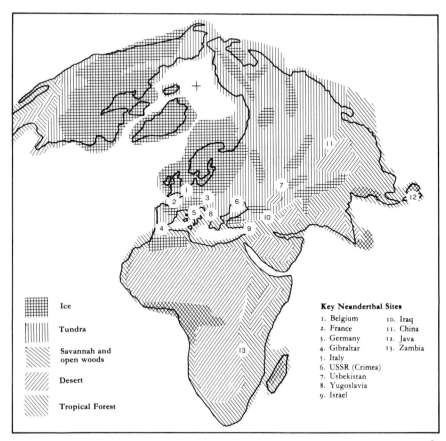

Ice

Tundra

Savannah and open woods

Desert

Tropical Forest

Key Neanderthal Sites

1. Belgium	10. Iraq
2. France	11. China
3. Germany	12. Java
4. Gibraltar	13. Zambia
5. Italy	
6. USSR (Crimea)	
7. Usbekistan	
8. Yugoslavia	
9. Israel	

Ill. 75. Map showing the distribution of Neanderthal man in Europe, Asia and Africa.

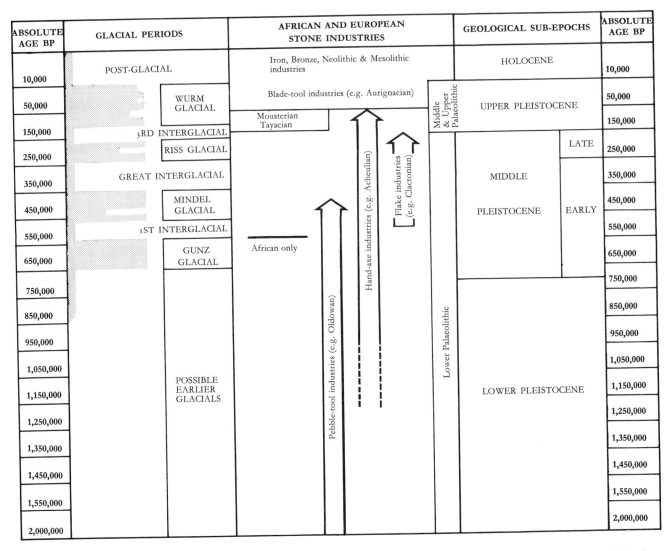

The Sabines, who took up arms against the Romans when their women were abducted, were ultimately absorbed by their conquerors and became Romans. Sabine genes became mixed with Roman genes. This sort of model provides a likely explanation for the apparent disappearance of the Neanderthalers. If this is true, then the Neanderthal race is still represented in the genetic bank account on which modern man still draws today. In flippant parenthesis it is probably permissible to assume that your 'nasty, brutish and short' next-door neighbour, who reminds you of a primitive caveman, might be just that—genetically speaking!

The first and second questions are not so easy. The answer to both involves a much bigger question. Where and when did modern man evolve?

The periods with which we are concerned are the Middle and Upper Pleistocene sub-epochs, which together span some 500,000 years. The chronology of the Pleistocene is particularly confusing as there are a number of different ways in which it may be subdivided. Geologists think in terms of the ages of rocks, glacialists in terms of Ice Ages, and anthropologists in

Ill. 76. A correlative scheme bringing together glacial, lithic and geological subdivisions of the Pleistocene epoch. Note that the origins of the pebble-tool industries extend back out of the diagram and into the Pliocene.

Ill. 76

terms of cultural periods. Each type of classification has its own advantages and limitations; for example, Ice Ages, which are widely used as Pleistocene chronological markers, are irrelevant for dating in the southern hemisphere. Africa had its rainy periods or pluvials but it is not really understood how these correlate with the alpine Ice Age divisions. The three most commonly used systems are shown in Ill. *76*, but the correlations between them must be regarded as very approximate.

In general terms, the Lower Pleistocene was the period of the more advanced australopithecines, including the hominids from Olduvai Gorge which have been designated as *Homo habilis* (see pp. 97, 98), and the earliest representatives of *Homo erectus*. The Middle Pleistocene saw the transition of men of the *erectus* grade to the *sapiens* grade which subsequently expanded geographically, giving rise to the four subspecies of man, including Neanderthal man, listed above. The following dates will act as a useful rule of thumb. It will be noted that as the Middle Pleistocene is so long, it has been arbitrarily divided into Early and Late sections.

Lower Pleistocene	2,000,000 to 700,000 years BP
Middle Pleistocene	*Early* 700,000 to 250,000 years BP
	Late 250,000 to 150,000 years BP
Upper Pleistocene	150,000 to 10,000 years BP

To say that chronological subdivisions are not important sounds somewhat offhand. They are, of course, very necessary adjuncts to communication between scientists; but more important are the absolute dates of particular fossils, because only in this way can any type of sequence or phylogeny be constructed. Fortunately there are two quite reliable methods available for dating fossils radiometrically: Carbon 14 (or C14) dating up to 60,000 years BP, and Potassium/Argon (or K/A) dating for rocks 250,000 years BP, or older. This leaves a tiresome gap between C14 and K/A, and here other methods, which do not give absolute dates, have to be employed.

The Middle Pleistocene Melting-pot. The earliest contenders for recognition as *Homo sapiens* are from Steinheim in Germany and from Swanscombe in Britain. Both skulls of Late Middle Pleistocene age are between 200,000 and 250,000 years old. Although both have been studied in the past by traditional methods of comparison, character by character, only recently have they been subjected to more advanced statistical techniques of multivariate analysis. Drs. Campbell and Weiner[26] took seventeen measurements from the Swanscombe skull bones and compared them with similar measurements taken from the skulls from Steinheim and Mount Carmel, from a number of Neanderthal skulls, and from a series of modern skulls from a Bronze Age population at Lachish in Palestine. The results demonstrated that Swanscombe was more primitive than was generally thought and lay about halfway between Neanderthal man and modern man. Steinheim was similar.

During the Late Middle Pleistocene and the Upper Pleistocene, different geographical populations of early *sapiens* were making contact in Asia, Europe and the Middle East. Probably this was the consequence of nomadism of hunting communities following in the wake of migrating herds. Whatever the reason, Europe and Asia were a kind of genetic 'melting-pot' in which groups of men, showing a wide range of physical variation,

Ill. 77. The Steinheim skull (female), from a gravel pit beside the River Murr, near Stuttgart, West Germany. It was discovered in 1933 by Karl Sigrist, jun.

reflecting the environmental differences of their places of origin, met and interbred. Out of the melting-pot emerged the most successful lines, which, during the Upper Pleistocene, resolved themselves into the four recognized subspecies of *Homo sapiens*. Even so, there must have been a considerable blurring between races, constituting a continuous spectrum of variation, rather than a series of discrete racial characteristics. An analogous situation exists with regard to the races of modern man, which, incidentally, have no connection with the races we are now talking about; modern races are geographical variants of *Homo sapiens sapiens* not of *Homo sapiens*.

Ill. 78. The Swanscombe skull, from the River Thames Gravels, 24 feet below the surface of the 100-foot terrace at Swanscombe, Kent. The original fragment was discovered by A. T. Marston in 1935 and subsequent pieces were found in 1936 and 1955.

The Swanscombe skull, somewhat modern, somewhat Neanderthal, is a true product of a melting-pot type of evolution. The two further examples, both from the Upper rather than the Middle Pleistocene, show that variants of *Homo sapiens* were not only sorting themselves out at this time, but were capable of co-existing in one area.

Round about 100,000 years BP, at a site near the Omo river in Southern Ethiopia, two individuals, rather impersonally known as Omo I and Omo II, died in unknown circumstances. This was no Capulet and Montagu situation, in spite of the build-up, for their skulls were found a mile apart and on opposite sides of the river! But, nevertheless, on geological grounds they are judged to be contemporary. Omo I is altogether of a modern type, including the presence of a chin, while Omo II is completely different. Professor Day,[6] who studied these skulls, considers that, in spite of its apparently large brain size, Omo II is an archaic form, having some similarities to *Homo erectus* but chiefly reminiscent of Rhodesian and Solo men, two of the four Middle-to-Upper Pleistocene races of *Homo sapiens*. There is no special reason to assume that these races (or subspecies) were evolving at the same rate, or that the various parts of their bodies were evolving towards a modern configuration in the same order. Mosaic evolution is the term used to describe the asynchronous development of physical characters within an evolving species.

Ill. 79. The Tabūn skull (extensively restored), from a cave floor at Mount Carmel, Israel. It was discovered by a joint expedition of the British School of Archaeology in Jerusalem and the American School of Prehistoric Research.

The second example comes from Mount Carmel in Israel, a site dated well on in the Upper Pleistocene. Two caves, both containing human remains, came to light in 1933 as a spin-off from Dorothy Garrod's archaeological work in the area, one at Tabūn and one at Skŭhl. The caves were occupied early in the Würm glaciation about 50–60,000 years BP. It is thought that the cave at Tabūn is older by about 10,000 years so, unlike Omo, we are not dealing with actual contemporaries—but what is 10,000 years between friends? Especially if they happen to be anthropologists.

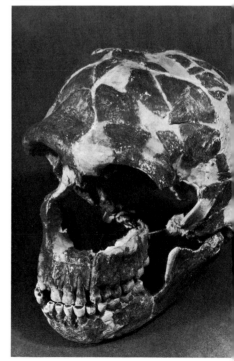

The Skŭhl skull is typically modern in aspect but Tabūn clearly belongs to the 'Classic' Neanderthals of the first part of the Würm glaciation, although it lacks some of their extreme characteristics, such as the flattened skull vault and the bun-shaped back to the head. One complete skeleton and several limb bones were also found, and these show that the Tabūn people were not so short and squat as their Ice Age successors.

Other discoveries of the Late Middle and the Upper Pleistocene at Shanidar, for instance, show Neanderthal-like characters. Here seven skeletons are known. There is also Amud man from Palestine, the latest find, who possessed a skull with a cranial capacity of 1740 cc—huge even by Neanderthal standards. Some skulls, such as those from Ehringsdorf and Saccapastore, are 'intermediate' in their characters; others, like Fontéchevade, are also intermediate but with a shift to the right which makes them rather more *sapiens*-like than Neanderthal-like.

There is thus plenty of evidence to support the melting-pot theory of the origin of *Homo sapiens sapiens* and *Homo sapiens neanderthalensis*. It is clear that Neanderthal man of the last Ice Age was not merely an isolated phenomenon, but rather the final expression of an ancient racial variant of *Homo sapiens*, identifiable in the fossil record of the Middle and Upper Pleistocene for at least 200,000 years.

Only part of the questions we set ourselves have been answered. The main one—where and when did *Homo sapiens* arise—has now to be tackled. In terms of the melting-pot analogy, where did the ingredients come from?

ERECTUS-SAPIENS INTERMEDIATES

One of the most puzzling discoveries of recent years has been part of a skull from Vertesszöllos, near Budapest in Hungary. It is puzzling because it possesses features that are remarkably modern in spite of being over 400,000 years old.

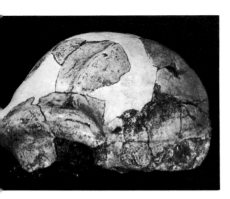

Ill. 80. The 'Chellean' skull cap (partly reconstructed) found by Louis Leakey in 1960 in Upper Bed II, Olduvai Gorge. The cranial capacity is quite large, about 1000 cc.

Its discoverer, Professor Andor Thoma, avers that the Vertesszöllos occipital bone shows a mixture of characteristics, some of which recall *Homo erectus*, and some which are more modern. Its brain size was large (1400 cc)—larger, in fact, than the *average* figure for modern man (1350 cc). Associated with the skull was a 'living floor', as it is called, on which stone tools of an Oldowan type (see below) and evidence of the use of fire were apparent.

It is worth noting that the Vertesszöllos bone comes from Central Europe, as does another Early Middle Pleistocene fossil, the jaw of Heidelberg man. The date of the Heidelberg jaw is very questionable as it was removed by a quarryman from a stratigraphical level that has never been positively identified. Based on faunal dating methods, the jaw is conceded to belong to the upper part of the First Interglacial (Early Middle Pleistocene), which gives it an age of approximately 400–450,000 years.

Scientists have found both these fossils difficult to pigeon-hole. Are they *Homo erectus* or *Homo sapiens*? Perhaps one day, when sufficient fossil material is available to indicate the range of variability, the matter will be resolved on a statistical basis; but in the meantime the only recourse is to refer to such intermediates non-committally—by their common names.

Ill. 80

It might be argued that since *Homo sapiens* came out of the West, like young Lochinvar, and *Homo erectus* came out of the East, the epicentre of evolution had shifted from one end of Eurasia to the other, and that Vertesszöllos was a kind of staging-post in the westward migration of the early men of China and South-east Asia. Against this view of the origin of European *sapiens* types is the fact that *Homo erectus* was already in existence in the West, although not in Europe but in Africa. He was present at Olduvai Gorge in the Lower Pleistocene, where the Chellean skull was discovered in Upper Bed II in 1960. *Homo erectus* was also present in North Africa at Ternifine. He may also have been present in South Africa at Swartkrans, where australopithecine sites have yielded up remains of an advanced hominid originally called *Telanthropus*. It is easy enough to accept that early men of the *sapiens* grade might have reached Europe from Africa, bringing their hand-axe culture with them. Acheulian hand-axes have recently been shown to be at least one million years old.[8]

By the Early Middle Pleistocene there existed two distinctive stone-tool industries, the Acheulian hand-axe industry of the West and the chopper-tool industry of the East, known as the Chou-k'ou-tien in China and the Early Soan in India. There are good reasons for deriving both the Chou-k'ou-tien-Soan and the Acheulian from the pebble-tool industry known as the Oldowan, which is now known to date back in East Africa for over two and a half million years. This geographical distinctiveness lasted until the Mindel-Riss Interglacial of the Early Middle Pleistocene, some 3–400,000 years ago.

Ills. 81–83

The discreteness of tool industries implies a lack of cultural interchange in *erectus* times which rapidly disappeared in early *sapiens* times. Further, the fact that the African *Homo erectus* followed the hand-axe tradition, while the Far East *erectus* populations did not, implies that the Asian and African *erectus* stocks evolved in geographical isolation.

From the evidence, the conclusion to be drawn is that early *Homo sapiens* was migrating eastwards, taking its particular (African derived) stone-tool culture with it. As there is no reason to doubt that the early eastern *sapiens* were expanding in a westwards direction, the two previously isolated populations would have met and interbred, thus initiating the world-wide unification of *Homo sapiens*.

It is all very well to talk about 'migrations' but early men didn't migrate for fun. There was no 'it's a fine day, let's migrate' spirit in the air. The incentive was hunting. For a population which owed its very existence to the killing of wild animals for food and domestic essentials, it was a matter of necessity to range further and further afield in search of game. There is a rather odious analogy to be drawn with the collectors of wild animals (for commerce not for food) today. Iquitos in Peru is a major centre for the collection of South American monkeys for ultimate sale to research laboratories in the United States. At one time, it was only necessary to range a few miles from the city to obtain adequate stocks; now the hunters have to travel a hundred miles or more from the city to maintain their shipments.

With these Palaeolithic hunters went their families, and in this manner the populations expanded westwards. Their progress would have been slow, perhaps a few hundred miles per generation. Migrations—one must remember—are not so much a matter of people-movement as of gene-movement.

Vertesszöllos could have been not a staging-point of Eastern early *sapiens* (née *erectus*) groups, but a meeting-point of East-West populations. It could have been the cradle of Wendell Wilkie's 'one world'.

Out of the post-*erectus* populations sprang a polytypic species, *Homo sapiens*, the melting-pot. From the melting-pot there emerged—as is always the way—a number of geographical variants. Some were successful, like modern man, some succumbed, like Neanderthal man, Solo man, and Rhodesian man, for a multiplicity of reasons that we may never understand.

HOMO ERECTUS

The Human Club has always operated on a closed-shop principle. To be elected the applicant is required to meet certain rigorous and precise criteria. According to one eminent British anthropologist, Sir Arthur Keith, he needed, for example, to possess a cranial capacity (or brain size) of at least 750 cc.

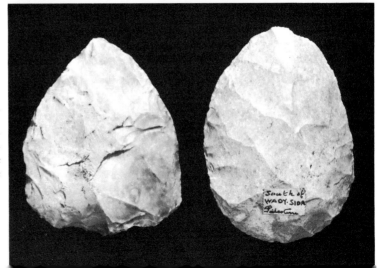

Ills. 81–83. Hand-axes formed an important item in early man's tool-kit for about one million years. These photographs demonstrate a gradual technical refinement in the construction of this all-purpose tool. Top: Lower Acheulian hand-axe dating from *c.* 500,000 years BP; centre: Middle Acheulian hand-axe, *c.* 250,000 years BP; bottom: Upper Acheulian hand-axe, *c.* 100,000 years BP.

In 1950 Dr. Ernst Mayr, the Harvard zoologist, felt that it was about time that human nomenclature fell in line with the principles operating in other fields of zoology.[14] He proposed to drop the endless list of generic names then current, and place all hominid fossils in the single genus *Homo*, on the grounds that most of these so-called genera represent, at best, species of a single genus and, at worst, geographical variants of a single species.

Mayr's attitude is a very fashionable one in modern taxonomy; it is known as 'lumping', which, in spite of its slightly pejorative ring, is a procedure that reflects current biological thinking on population dynamics. Reactionaries—splitters—are severely frowned upon for their typological attitudes.

Pithecanthropus (Java and Peking man), *Atlanthropus* (Ternifine man) and *Ill. 84* Chellean man from Olduvai have been 'lumped' into the species *Homo erectus*. Dubois, the discoverer of Java man, called his discovery *Pithecanthropus erectus*. The trivial name, *erectus*, has been kept, so *Homo erectus* it is.

Eugene Dubois was a Dutchman who entered military service for the express purpose of serving in the Dutch East Indies where, he reasoned, the origins of man were to be found. It seems almost incredible that in 1890, when only a few Neanderthal skulls were known from Europe, anyone should have had the self-confidence to predict that the origins of man lay not in Africa, as Darwin had averred, but in South-east Asia. No-one really knows why Dubois was so ready to back his hunch but it is not unreasonable to suppose that he was determined to out-guess Darwin.

Darwin's prophecy was based on the presence of the chimpanzee and gorilla in Africa. Perhaps Dubois considered that the orang-utan and the gibbon of the Dutch East Indies had greater claims to human ancestry. Perhaps, too, he had observed, as Sir Arthur Keith did in 1924, that the upright posture of the brachiating gibbon was functionally similar to that of bipedal man. That the gibbon was in Dubois' mind is suggested by his statement in 1891, following the first discovery of the skull of Java man, that it 'could be compared with that of a gibbon enlarged to twice its size'. Whatever his motives, his actions were crowned with success. After barely a year he had discovered a skull cap and a femur on the banks of the Solo river near Trinil on the island of Java. Scientific reaction was predictable and uniformly scathing: it was 'a primitive orang-utan', 'an extinct giant gibbon', 'a microcephalic idiot' (Neanderthal man, if you remember, was labelled by some as a macrocephalic idiot. Fossil hunters can't win!), and finally, it was 'a monstrous offspring of human and simian parents'.

Dubois behaved rather oddly after his first magnificent discovery. Embittered by the criticisms of anthropologists, he kept his mouth shut for the next thirty years and only in 1924 did he reveal that in 1891 he had also discovered the lower jaw of Java man. The jaw was a bit of an eye-opener and helped considerably to elevate the status of *Pithecanthropus*, as it was then called, from a near-ape to a near-man.

- Dubois' compatriot and *de facto* successor, Professor von Koenigswald, concentrated his excavations on the Djetis Beds which underlie the Trinil Beds where Dubois' specimen was found. Between 1937 and 1939 von Koenigswald unearthed parts of four different skulls, one that of a child, and a mandible bearing what seemed at that time to be exceptionally large molar and premolar teeth. Compared with such Lower Pleistocene forms as *Paranthropus*, the robust australopithecine from South and East Africa, the

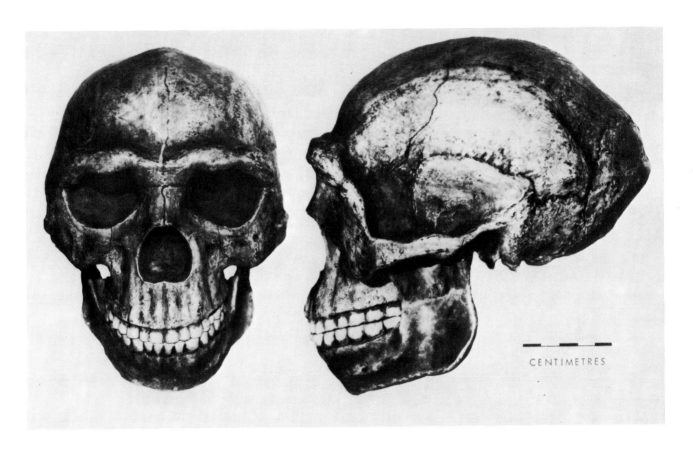

Ill. 84. Cast of the restored skull of Peking man (*Homo erectus*) from Chou-k'ou-tien, China.

teeth are not particularly impressive. However, von Koenigswald named this jaw *Meganthropus*. It seems likely that *Meganthropus* was an australopithecine which spread out of Africa two million years ago and was on its way to becoming *Homo erectus*.

The dating of the Java material is uncertain. The Trinil zone probably lies between 500,000 and 800,000 years old, thus spanning the Lower Pleistocene and Middle Pleistocene sub-epochs. The Djetis zone is older, lying anywhere between one and two million years BP, though probably nearer the former date than the latter.

By the mid-1920s, Java man's probable descendant, Peking man, had been discovered in limestone caves at Chou-k'ou-tien, north of Peking, by an international team of palaeontologists funded by the Rockefeller Foundation. Between 1927, when the first skull came to light, and 1941, when the Japanese attack on Pearl Harbour brought excavations to an end, a dozen skulls, several jaws, numerous isolated teeth and a few limb bones had been found. At the outbreak of war they were packed and despatched by ship to America. They have never been heard of since. What happened to them is a total mystery. Fortunately most of the originals had been photographed and casts had been made.

Peking man (now *Homo erectus*) was rather short of stature and stockily built, unlike his more rangy tropical forerunners from Java. He was a hunter and a tool-maker; he had discovered the use of fire as a means of keeping *Ill. 198* warm, of providing light, of cooking food, and of frightening off dangerous predators like the ubiquitous cave bears. The discovery of fire permitted man for the first time in his history to spread outside the tropics. Chou-k'ou-tien lies on the 40°N parallel; the present climate offers dry, cold winters and hot, wet summers, but in Middle Pleistocene times the climate was considerably cooler. From the evidence of fossil vegetation it was almost boreal in nature. Thus it is easy to understand Peking man's rather stocky build; in fact, he was the prototype of later European races, like the Neanderthalers to take an extreme example, whose body form was adapted for heat conservation. Anatomically Peking man was already on the way towards the metamorphosis of *Homo erectus* into *Homo sapiens*. His brain size (Av. 1000 cc) lies at the lower end of the size scale of modern man (Av. 1350 cc), but was in advance of Java man (Av. 940 cc). His skull was thick-walled and flattened but, again, not so flat as Java man's. His brow ridges were excessively well developed and the lower part of his face was somewhat protrusive, a consequence, no doubt, of his relatively large teeth. The upper canines were stout and projected downwards to reach slightly below the level of the remainder; the lower canines were unremarkable.

The age of the *Homo erectus* remains from the Chou-k'ou-tien caves is difficult to establish as all the fossil sites were infillings in limestone caves. It seems certain that the caves were occupied during the Early Middle Pleistocene, probably during the time of the Mindel Glaciation 400,000 years ago. A preliminary estimate of 300,000 years has recently been given by Drs. Bada and Protsch, based on a newly-developed dating technique using an amino acid reacimization reaction.[1]

The presence of *Homo erectus* in Africa has already been discussed, and a new skull, not yet fully described, from Petralona in Greece may also prove to belong to the *erectus* species. There is only one more claimant for *erectus* status and a very enigmatic one it is—or rather 'they' are since the discovery of Solo man consisted of eleven skulls. Solo man was found at Ngandong on the Solo river, Java, in deposits of somewhat more recent date than the Djetis Beds. Once again, the dating is uncertain but it is probably Upper Pleistocene and therefore less than 150,000 years old. As regards the affinities of Solo man, opinions alternate between Neanderthal-like and *erectus*-like. In Louis Leakey's view, Solo man was a late survivor of Java man. This is not the consensus opinion, however, which classifies Solo man as *Homo sapiens soloensis*, although without very good reason, let it be said.

THE AUSTRALOPITHECINES

It is a measure of the present healthy state of human palaeontology that the status of so many 'old faithfuls' is being challenged. The case of the Taung child is a good example.

The skull from Taung was the original discovery that led to the australopithecine bonanza of the late 1940s and 1950s. Although the skull was found in Bechuanaland by Raymond Dart in 1924, it elicited little enthusiasm from anthropologists, who were still obsessed with Piltdown man. After all, what chance did a fossil from the colonies (which didn't even look like a man) have, compared with the magnificent specimen from Sussex-by-the-sea? The little australopithecine from Taung, ultimately given the name *Australopithecus africanus*, was brushed aside (shades of the Neander Valley man and Java man!) as a fossil ape.

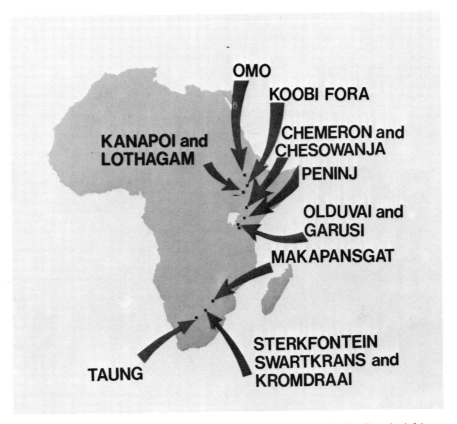

Ill. 85. Sites in South and East Africa where Lower Pleistocene hominids have been discovered.

Taung has always been looked upon as the oldest site in the South African sequence, but in the last couple of years it has slid—as in the old traditional nursery game—all the way down a snake to become the youngest of all. Based on Dr. Partridge's newest dating of South African cave sites[20] (an almost impossibly difficult task), Taung has been relegated from a respectable two to three-million-year antiquity to something less than one million years, a conclusion that has been received with undisguised scepticism by most authorities.

The point however is, whether right or wrong, that the traditional view has been challenged and up to a point this can only be a good thing for anthropology. Nature can be challenging too, as the new discovery near Lake Rudolf of the skull of a human being living, apparently, a million years before his time, clearly demonstrates (see p. 99).

The 1970s are proving a good decade for physical anthropologists. Once it was a question of too many scientists chasing too few fossils, but now it is a matter of finding enough properly trained people to do the work which is accumulating at an unprecedented rate.

Ill. 92 The australopithecines are a group (taxonomically a subfamily) which lived in Africa from the beginning of the Pliocene epoch to the middle of the Pleistocene, a time-span of five to six million years. The earlier representatives of this prehuman grade were first identified in South Africa but, as we now know, the subfamily was widely distributed in South and East *Ill. 85* Africa from the Transvaal, Tanzania, Kenya, and southern Ethiopia. The earliest australopithecines known come from Lothagam Hill, to the west of

91

Lake Rudolf in the Northern Frontier district of Kenya. The evidence from Lothagam, one molar tooth embedded in a fragment of a lower jaw, is miniscule but it is enough to make us revise our previous estimates of the age of this group.

What was once thought to be a relict, aberrant, by-blow of hominid evolution tucked away in the southern redoubt of a continent, is turning out to be a successful population of near humans, widely dispersed in time and space. Time has been kind to the australopithecines, but to the once much-revered Piltdown man it has been harsh and unrelenting.

South African Australopithecines. According to one's tendencies for 'lumping' or 'splitting', the South African australopithecines consist of either one genus with two species: *Australopithecus africanus* and *Australopithecus robustus*; or two genera each with a single species: *Australopithecus africanus* and *Paranthropus robustus*.

In order to simplify the following account and at the same time gracefully side-step all taint of taxonomic in-fighting, I propose to follow the convention of referring to the two forms as the gracile and the robust. The colloquialisms are appropriate, for *Australopithecus africanus* is lightly built in all its known features, while *Australopithecus robustus* (or *Paranthropus*) is somewhat burlier.

Ills. 86, 87

The gracile form is well known from a large collection of fossils (parts of sixty-three individuals) from three main sites in South Africa: Makapan, Sterkfontein and Taung. The remains include several more-or-less complete skulls, four hip bones, bones of the vertebral column and of the arm, the hand, the leg, and the foot. The robust material comes from Swartkrans and Kromdraai and consists of parts of ninety-four separate individuals.

Collectively, this evidence provides us with an anatomical profile of gracile australopithecines of which the following are the principal features: (1) Small brain case with large jaws and teeth. Brain volumes according to the most recent estimates average 442 cc. This compares unfavourably with the mean for adult gorilla skulls of 506 cc, but it must be remembered that brain size and body size are closely correlated. Gorillas are considerably larger animals than australopithecines, whose height has been estimated by some workers at 3 ft. 6 in. (2) Upright posture with head balanced in a near man-like fashion on the top of the vertebral column. (3) Moderate protrusion (prognathism) of the face is intermediate in degree between modern man and modern apes like the chimpanzee. (4) Teeth, large compared with those of modern man, but small compared with those of the apes. This is particularly true of the canine teeth, which do not protrude beyond the tooth row, and are small in size and incisor-like, as they are in man. The pattern of tooth wear, reflecting the movements of one jaw upon the other in chewing, is man-like; in apes the large canines tend to limit the side-to-side chewing movements which produce evenly ground-down surfaces on the molars. (5) The walking gait was habitually bipedal but less well executed than in modern man. The quintessence of the human walking gait is the ability to stride, as I pointed out first in 1963[15] and later discussed more fully in 1967.[16] Striding is the most economical way to deploy the anatomical and physiological attributes of the human body in the context of the hunting behaviour of man that played such a critical role in his later evolutionary stages. The gracile

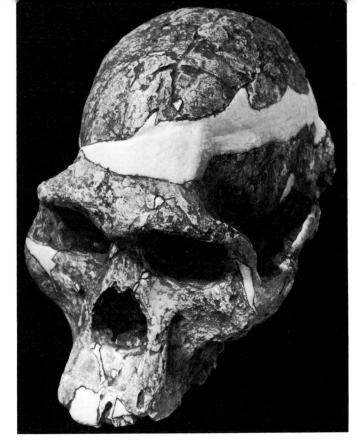 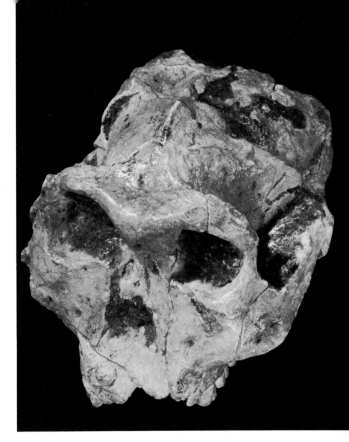

Ills. 86. 87. Australopithecine skulls. Left, the gracile type (*Australopithecus africanus*) and right, the robust type (*Australopithecus robustus = Paranthropus robustus*).

australopithecine was not a strider in the sense that *Homo habilis* probably was, although he was a capable bipedal walker.

The robust species differs in a number of important ways from its gracile cousin. Its body size was greater: it stood between 4 ft. 10 in. and 5 ft. 4 in. in its socks. Its brain size, consequently, was also greater, about 530 cc. The jaws were considerably more robust, housing massive molar teeth, and the skull bore the evidence of the attachment of strong jaw muscles needed to operate them. The teeth lacked the prominent canines of the apes (as did those of the gracile form) but differed from the gracile form in a quite remarkable way—the front teeth (incisors and canines) of the robust form were proportionately much smaller and the cheek teeth (premolars and molars) proportionately larger.

This particular size relationship appears to be an adaptation to diet and is found in certain non-human primates, notably an Old World monkey, the gelada, a baboon-like creature found at high altitudes in central Ethiopia. Geladas feed on montane moorland at altitudes where the grasses have become sparse and tussocky. They have evolved a herbivorous type of diet (known as graminivorous) in which grasses, seeds, rhizomes and bulbs form the major elements. These food items are both small and tough and require extensive crushing and chewing. This feeding habit led to the evolution of large molars and premolars, which are packed closely against one another, forming, in effect, a solid, uninterrupted bar of dentine and enamel. As food items are collected by hand, the front teeth play little part in this type of feeding and consequently are relatively small.

The shape and function of teeth have a great deal to do with the shape of the face. Geladas possess rather flat faces, deep from top to bottom and

93

heavily muscled round the angle of the lower jaw. A parallel has been drawn by Dr. Clifford Jolly between the geladas' facio-dental complex and that of the robust australopithecines.[9] Undoubtedly australopithecines were ground feeders and their diets would have included much the same components as those of the geladas. Although the gracile forms show something of the same facio-dental tendencies as the robust, the conclusion to be drawn is that the ecology of the robust and gracile australopithecines was markedly different.

Finally, there is the matter of gait. Although the robust type was capable of two-footed walking to a degree more advanced than the great apes, its walk was notably less perfected than the bipedalism shown by the gracile type, and a far cry indeed from the effortless stride of modern man.

A rather interesting exercise in demography has been carried out recently by Professor Phillip Tobias.[25] He has demonstrated that when age-identifiable fossil specimens from the five South African cave sites are compared, it becomes apparent that there is a twenty-five per cent greater mortality

Ills. 88, 89. Left: part of the skull of a young australopithecine from Swartkrans showing holes which are thought to have been made by the canine teeth of a Lower Pleistocene leopard. The lower jaw of such a leopard is held in position against the skull, showing that the canines fit exactly into the puncture holes. The reconstruction (below) illustrates how the damage to the skull could have been caused by a leopard dragging the dead child to its feeding place.

Ill. 90. The lower Gorge at Olduvai, looking south to the extinct volcanos of Sandeman and Lemagrut across the Serengeti Plain. The basal basalt is visible at the bottom of the Gorge. Overlying this are the horizontal bedded volcanic tuffs and lacustrine deposits of Beds I and II. The Upper Beds are the red sand dunes of Beds III and IV. On the left of the picture is one of the main faults.

among immatures of the robust group than the gracile group. Obviously this suggests that some ecological factor was operating at Swartkrans, where the majority of robust australopithecines come from, and not at Sterkfontein or Makapan, which are 'gracile' sites.

The work of Dr. C. K. Brain, of the Transvaal Museum, may be relevant to this observation.[3] Brain has been studying the behaviour of leopards in an attempt to explain why australopithecine bones accumulated in cave sites and, further, why the accumulations were so rich in skulls and jaws and so poor in postcranial bones. Carrying out feeding experiments with living cheetahs (used simply as convenient models rather than as suspects), he determined that their preference is for the relatively fragile parts of the animal skeletons. For example, they would consume the backbone of a baboon but leave the denser parts at the extremities of the limb bones relatively undamaged. The hands and feet were always eaten, the skulls were usually untouched.

These findings go a long way to explain why some parts of the skeleton are common in the cave accumulations and some are very rare, but they throw little light on the preponderance of immature individuals found in fossil deposits.

In 1949 a robust skull (SK 54) of a young adult was found at Swartkrans. *Ill. 88* On the back of the skull two puncture holes, 33 mm. apart, were apparent. Brain carried out a series of experiments with the jaws of living and extinct South African carnivores and found that the canines of the lower jaw of male *Ill. 89* and female leopards varied between 30 and 41 mm. and that the mean width in a series of female leopards was 33 mm.! Lower Pleistocene leopards were

95

smaller than their modern counterparts and, although it is possible that adult australopithecines could have been dragged away with their heads gripped firmly in the leopard's jaws (allowing the body to trail between the legs), there can be no doubt that a young specimen would have been much easier to handle.

The greater susceptibility of the robust juveniles to an early demise (by leopards or, as some popular writers would hold, by internecine killing—sometimes anachronistically labelled as 'murder') might have another explanation. The evidence from South African sites is that tool-making was not nearly as advanced as it was amongst later australopithecines at Olduvai; in fact, it is uncertain whether australopithecines made tools in the accepted sense at all. However, it is a fair assumption that if they did not make them, they certainly used them. If they were not of stone they may well have been of bone. A bone-tool culture in which bones were used with a minimum of adaptation has been proposed with the jaw-cracking name of the Osteodontokeratic industry. Of the two australopithecine species, or genera as I would prefer to regard them, the anatomical evidence suggests that the graciles were more man-like than the robusts. This being so, it is likely that the graciles were the more adept in all departments (walking, tool-use, etc.) than the robusts. Furthermore, their teeth suggest that they could have been partly flesh-eating and, thus, more capable and experienced as hunters. In these circumstances, even the juveniles would be less likely to fall prey to a marauding leopard than their less advanced, less agile and more robust cousins.

East African Australopithecines. Olduvai Gorge is strictly 'Leakey country'. The discoveries of the Leakey family at Olduvai during the 1960s led directly to the present healthy state of African palaeoanthropology. Whatever and wherever future discoveries are made, anthropologists will inevitably come to look upon Olduvai as the cradle of the new era.

Olduvai Gorge was first discovered by Professor Kattwinkel in 1911 and has been recognized as a fossil site of great interest since 1913 when, just before the outbreak of the First World War, Professor Hans Reck discovered a fairly modern human skeleton which came to be called 'Oldoway man'. Although the potential of the Gorge was known, it wasn't until 1931 that Louis Leakey started to work in the area. Between 1931 and 1959 indirect evidence of human occupation turned up in abundance in the shape of stone tools of a more primitive form than any previously recognized; but no direct evidence of human fossils came to light until 1959. In that year, Mary Leakey spotted an upper jaw partly buried in the soil of Bed I, and her first remark must go down as one of the most instantaneous christenings of all time. 'Ah! the dear boy', she said, and 'Dear Boy' it has remained in the family circle. Actually, no fossil hominid can have been christened so often. The press christened it 'Nutcracker man' on account of its enormous molar teeth, Leakey himself christened it more formally *Zinjanthropus boisei*, and less formally 'Zinj'. Its obvious relationship to the robust australopithecine of South Africa necessitated a change to *Australopithecus boisei*. Finally, for those who do not consider that robust australopithecines belong to the genus *Australopithecus*, 'Dear Boy' is *Paranthropus boisei*. But whatever it is called is of less importance than what it is.

Ill. 92. Opposite: an imaginative reconstruction of *Australopithecus africanus* in its natural habitat.

Ill. 90

Ill. 91. The skull of *Zinjanthropus boisei*, from Bed I, Olduvai Gorge. The lower jaw is hypothetical.

M.Wilson 1950

Ill. 93. Opposite: for many millennia men gathered wild cereals in the Ancient Near East, slowly removing some of the pressures of natural selection and so favouring mutants more suited to his needs. On the right: husked emmer. Each grain is encased, which helps natural propagation but makes threshing difficult. On the left: grains of naked bread wheat. Here several mutations have occurred and the husk has entirely disappeared. (See p. 118.)

The fame of 'Zinj' was soon eclipsed by an even more important discovery in the same general area of the Gorge. Early in 1961, at a very slightly lower geological level a few hundred yards away, the skull, lower jaw, foot bones, hand bones, and a collar bone of a young hominid came to light. Provisionally the newcomer was called 'pre-Zinj'. It soon became clear that 'pre-Zinj' was very different from 'Zinj'. 'Zinj' had massive molars, a heavy buttressed skull and a brain size of 530 cc. 'Pre-Zinj', on the other hand, had relatively small molars, a lightly built skull and a brain size of 657 cc.

In the following years, further specimens of 'pre-Zinj' were found in Bed I and in the lower part of Bed II below what is known as the 'faunal break'.

The two forms in Bed I appear to be in much the same relationship with each other as the gracile and robust forms of South Africa—with one major difference. The two Oldowan forms lived side-by-side both in time and space. Indeed, both 'Zinj' and 'pre-Zinj' sites contained bones of *both* forms. Such contemporaniety is perhaps the best argument in favour of splitting the two forms into separate genera—*Australopithecus* for the gracile and *Paranthropus* for the robust. Were they species of the *same* genus, one would have expected them to have been so strongly competitive in terms of food requirements that they could not have co-existed. The putative differences in their diet and behaviour, as reflected in their disparate morphology, suggests the distinctiveness of their life-styles. A model for this situation is seen today in the Gombe Stream Reserve, Tanzania, where chimpanzees (*Pan*) and baboons (*Papio*) co-exist. Chimpanzees occasionally predate on young baboons, just as *Australopithecus* may have done on *Paranthropus*, but such events are not incompatible with a state of peaceful co-existence.

The discovery of pebble-choppers, primitive Oldowan tools, at the 'Zinj' site led to 'Zinj' being canonized as the earliest human tool-maker, an honour that was very rapidly transferred to 'pre-Zinj'. The issue is of academic interest only since, as we now know, the Oldowan culture evolved at least one million years earlier than Bed I, Olduvai. At the time, however, it was of special interest since it gave point to the formal designation—in 1964—of 'pre-Zinj' as *Homo habilis,* Man-the-tool-maker. From the start, *Homo habilis* was an unpopular member of the Human Club; his election was considered to be precipitate. It was not disputed that *Homo habilis* represented an advanced member of the genus *Australopithecus,* but rather that his morphology was not sufficiently distinct from known representatives of the genus to merit such arbitrary elevation to the *Homo* grade. To the ascribers of *Homo habilis* (Louis Leakey, Phillip Tobias and myself[11]), the persuasive facts were the relatively large size of the brain, a lessening of the disproportion between the front teeth and the cheek teeth, the evidence derived from the foot bones that *Homo habilis* walked upright on two legs much as modern man does and, finally, its association with tool-making. These seemed reason enough at the time.

The '*habilis* controversy' continues, but it is of less moment today since the emotive component—that of being the earliest species of true man—has been overtaken by events. It now seems likely that Richard Leakey's skull from East Rudolf, KNM-ER 1470, has seniority.

Ill. 94

Since 'pre-Zinj' several other skulls attributed to *Homo habilis* have been found at Olduvai in Bed I and the lower part of Bed II, notably a skull from the oldest levels of Bed I, affectionately known as 'Twiggy' because, when

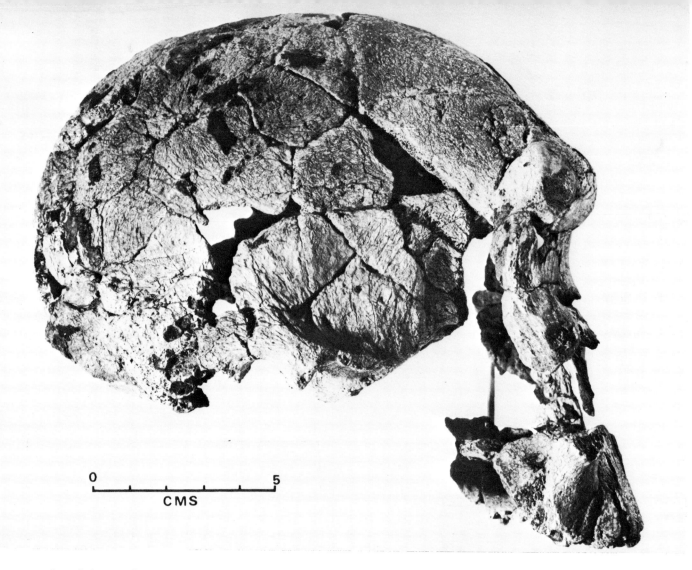

found, it was flat and lacked natural curves. This skull, plus 'George' from lower Bed II and 'Cinderella' from higher in Bed II above the faunal break, have led to a revision of the range and mean of the brain size of the species: range: 520–700 cc, mean: 657 cc. The time-span during which *habilis* occupied the site of the present Gorge has been extended by these discoveries to about one million years. However, recent discoveries elsewhere in East Africa, notably East Rudolf, have led to a reappraisal of the Bed I and II habilines. 'Twiggy', in particular, is considered to be more australopithecine-like than habiline.

Ill. 94. A provisional reconstruction of East Rudolf man—KNM-ER 1470.

One of the principal difficulties besetting students of African pre-history is the correlation of dates at different sites. The techniques applied to the East African sites cannot be applied to South African cave deposits which are due to in-fillings of previously existing solution cavities in limestone rocks. Isotopic dating by the Potassium/Argon technique has provided firm dates of the absolute age of rocks at Olduvai, such as those in Bed I associated with 'Zinj' and 'pre-Zinj', of 1·75 million years. On the basis of faunal and geomorphological dating, which by its nature cannot be 'absolute', the gracile australopithecine caves at Makapan appear to be around 3·7 million years old, and those at Sterkfontein, 3.3 million. On the basis of these figures it might be supposed that there was time enough for *Australopithecus africanus* to have evolved into *Homo habilis* at Olduvai. As we shall see, some of the other East African sites are older than the South African ones and some are younger. The critical site for our present purposes is below the KBS tuff at Koobi Fora, East Rudolf. Supposing that KNM-ER 1470 turns out, as well it may, to be a forerunner of *Homo habilis* from Olduvai, the sequence, *Australopithecus africanus* (3.7–3.3 million)–KNM-ER 1470 (3.1 million)–*Homo habilis* (1.75 million) appears plausible on chronological grounds.

Other East African Australopithecines and Hominines. Since 1967 several new areas, supremely rewarding for fossil hunting, have been opened up in northern Kenya and southern Ethiopia. The most important of these are at Lake Rudolf (Kenya), Omo Valley (Ethiopia), Kanapoi and Lothagam (Kenya). Unlike the South African fossils which have to be drilled out of consolidated breccia, the East African material, eroded out of sedimentary beds laid down in sequence, are waiting to be picked up. Consequently, not only are the fossils easier to date but they occur in greater abundance and are in better condition.

The earliest sites that have yielded hominid remains are at Kanapoi (4 million years) and Lothagam (5–6 million years). Both specimens are more like the gracile than the robust australopithecine. Apart from these two early examples, the evidence for graciles in East Africa has not been forthcoming. However, Richard Leakey has recently reported[13] a skull from East Rudolf (KNM-ER 1813) which appears to be of the gracile type. The two most commonly occurring forms are *Australopithecus robustus* (or *Paranthropus robustus*) and an indeterminate species of *Homo* known euphemistically as *Homo sp.*

During 1972, the Kenya National Museum's field expedition to Lake Rudolf in the Northern Frontier district of Kenya, under the direction of Richard Leakey, collected a number of important specimens that seem at first sight to be referable to the genus *Homo*. Attention, so far, has been focused principally on the skull known (with admirable restraint) simply as KNM-ER 1470. The remarkable thing about Richard Leakey's discovery is its age. The deposits are well dated at about 3 million years BP. Should it prove to be a member of the Human Club it would be older by more than a million years than the suspect *Homo habilis*, and by at least 2 million years than the first unequivocally human species, *Homo erectus*.

1470, as he is known to his friends, consists of a brain case, part of the face, and part of the upper and lower jaws. Rough estimates of brain size indicate a figure of the order of 800 cc, which is broadly equivalent to that of *Homo erectus* from Java which lived $2\frac{1}{2}$ million years later. Additionally, three femurs were found in association with the skull; the best preserved shows some similarities with the femur of modern man, combined with some *erectus*-like features known from Dubois' original discovery in Java.

A full report is still awaited on this very exciting specimen, which could make our present views of human evolution as out-dated as a crystal set. We can only speculate on the possible outcome, but speculation, providing it is recognized as such, is the life-blood of science. So, assuming that the age estimates of 1470 are correct, it is possible to visualize three possibilities: (1)

99

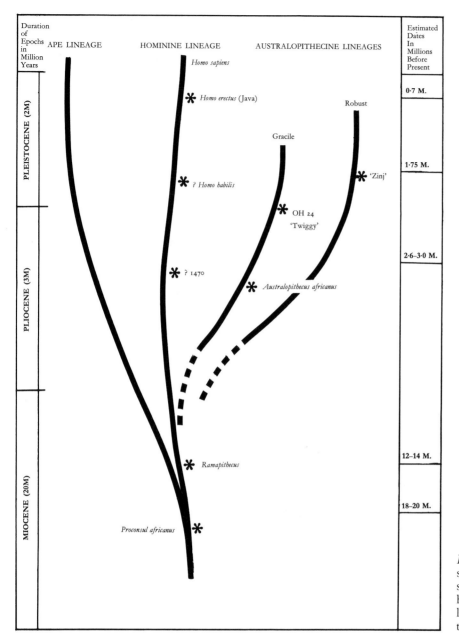

| Duration of Epochs in Million Years | APE LINEAGE | HOMININE LINEAGE | AUSTRALOPITHECINE LINEAGES | Estimated Dates In Millions Before Present |

Homo sapiens

✳ *Homo erectus* (Java)

Robust

Gracile

0·7 M.

✳ *? Homo habilis*

1·75 M.

✳ 'Zinj'

✳ OH 24 'Twiggy'

2·6–3·0 M.

✳ *? 1470*

✳ *Australopithecus africanus*

PLIOCENE (3M)

✳ *Ramapithecus*

12–14 M.

18–20 M.

Proconsul africanus ✳

MIOCENE (20M)

Ill. 95. A scheme showing one possible interpretation of the relationships between the african ape, the hominine and the australopithecine lineages. No attempt has been made to include all known fossils.

that *Homo sapiens* differentiated at a much earlier date than supposed; in which case, we would have to accept the *erectus* grade as an incidental side-branch of no great consequence. (2) that the *erectus* lineage was already established in the Upper Pliocene, 3 million years ago. (3) that KNM-ER 1470 is an early representative of the Lower Pleistocene species, *Homo habilis*, one branch, perhaps, of the earliest australopithecine stock.

For better or for worse, my money is on Option (3) and in this view I am not alone. It has already been espoused by Professor P. V. Tobias of Witwatersrand University[25] and by Dr. W. W. Howells of Harvard.[17] Perhaps by the time this book is published I shall be wishing I had kept my council.

If Option (3) turns out to be correct, then the Oldowan habilines will fall into place as later members of the same species. This is not too outlandish an idea, since both were tool-makers working in the same cultural tradition. At the nearby site at Koobi Fora, 'Oldowan' stone tools, dating back 3 million years, have been reported on by Mary Leakey.[12] Turning to anatomical features, both provide evidence that upright bipedal walking was well established and both possess quite large brains. Theoretically the discrepancy between the upper limit for the habilines of 700 cc and the estimated 800 cc for 1470 need not be an insuperable objection if the Oldowan specimens were juveniles and females and 1470 was a male.[7] In fact, all four skulls attributed to *Homo habilis* appear to be either females, juveniles or young adults. It will be many years before the information, presently available, is sorted out and a meaningful Plio-Pleistocene hominid lineage constructed. Paradoxically, the more fossil material that is discovered, the more difficult the task of unravelling the skein of human evolution becomes. It would be most unwise at this juncture to theorize too much. However, it can do no harm to summarize what we know in the broadest of terms.

The hominid story starts with the separation of the ape and human stocks discussed above. We have no precise date for this but it is assumed to have taken place some 15–20 million years ago. The earliest putative hominid that we are aware of is judged to have been *Ramapithecus*. Subsequently, the hominid stock underwent an adaptive radiation, a familiar enough phenomenon in evolutionary zoology, in which a new and successful species branches out into a number of derived species each adapted to a particular environmental niche. This event must have occurred about 10 million years ago. Judging by the 'end products', at least three distinct hominid lineages (species in the first instance) must have arisen during the Upper Miocene: (1) A hominine lineage leading ultimately through KNM-ER 1470, *Homo habilis* and *Homo erectus* to *Homo sapiens*. (2) An australopithecine lineage leading to the gracile *Australopithecus* of South and East Africa. (3) A second australopithecine lineage leading to the robust australopithecine, which I would prefer to call *Paranthropus*.

In *Ill. 95* these three lineages are shown as evolving at much the same time. The suggested inter-relationship of these is *theory*, but what is *fact* is that three such lineages co-existed in Africa 2·5 to 3·0 million years ago.*

The nature of the environmental forces that prompted this triple radiation is not understood, but in the case of the hominines it is likely to have had something to do with the development of culture in the form of tool-using and tool-making. In the case of the australopithecines, the principal factor that separated them may well have been environmental. The anatomical differences between *Australopithecus* and *Paranthropus* suggest that they followed quite distinctive ways of life in terms of habitat and diet.

RAMAPITHECUS

It is quite a journey back in time from the earliest known australopithecine at Lothagam to *Ramapithecus*, the first species to display hominid characters—some eight million years in fact.

There are two species of *Ramapithecus*; one from Fort Ternan, Kenya, *Ramapithecus wickeri*, and one from the Siwalik Hills in Northern India, *Ramapithecus punjabicus*. The African species, at about fourteen million years,

* While this chapter was in press a new discovery was announced by Dr. Karl Johanson and Dr. Maurice Taieb. These anthropologists claim that they have found the upper and lower jaws of a *Homo*-like creature in Eastern Ethiopia near the Awash River. The point of special interest about this discovery is that provisional dating has placed it at 1·5 million years *older* than KNM-ER 1470; at about 4 million years BP in fact. Obviously if the claims prove justified, we shall have to revise our views on the date of the origin of true man.

is slightly older than the Indian form at twelve million years. Unfortunately material of both species is limited to upper and lower jaws and parts of a lower face, but jaws and teeth, as we have seen, can be very helpful in providing information about the diet and way of life of the owners.

The principal human characteristics of the teeth of *Ramapithecus* are the small size and lack of prominence of the canines, the broad flat molars showing interstitial wear, and relatively small incisors. These characters are in striking contrast to the teeth of the contemporary dryopithecine apes with their huge, overlapping canines, large shovel-shaped incisors and a distinctive wear pattern.

The lower face indicates that the jaws were not protruded into a muzzle, as in the apes, but flattened as in australopithecines, particularly the robust forms. The facio-dental complex shown by *Ramapithecus* suggests that it was a ground feeder dependent for survival on such food items as grasses, seeds, rhizomes, bulbs and fleshy plants—quite a different menu from the fruit diet of the dryopithecines. The adaptive changes associated with graminivorous feeding have already been considered with reference to the robust australopithecine and the living gelada baboon (see pp. 93, 94).

From the evidence provided by fossil seeds and the remains of other mammals of the Nagri Formation of the Siwalik area, it seems that *Ramapithecus* lived in a region of mixed forest interspersed with open areas around lakes and watercourses. The situation in East Africa for *Ramapithecus wickeri* was much the same except for the presence of rather more grassland. This type of environment is very similar to that proposed by me in 1967 as the ideal nursery for evolving hominids:[16]

> An environment neglected by scholars but one far better suited for the origin of man is the woodland-savannah, which is neither high forest nor open grassland. Today this half-way-house niche is occupied by many primates, for example the baboons, the vervet monkeys and some chimpanzees. It has enough trees to provide both forest foods and ready escape from predators. At the same time its open grassy spaces are arenas in which new locomotor adaptations can be practised and new foods can be sampled. In short, the woodland-savannah provides an ideal nursery for evolving hominids, combining the challenge and incentive of the open grassland with much of the security of the forest. It was probably in this transitional environment that man's ancestors learnt to walk on two legs. In all likelihood, however, they only learnt to stride when they later moved into the open savannah.

CONCLUSION

In chapters 3 and 4 we have looked at human evolution from back to front and from front to back. All the important facts available to us have been given. They are a drop in the ocean. There are millions of unknown facts to be discovered before the script of our drama is complete. At present we have to be satisfied with the rough-edited, bitty, and inconclusive style of the *cinéma verité*.

One thing, I believe, emerges. Man has been on the cards since the primate order first took shape. Every step, every biological improvement that has characterized the evolution of the primates through time has been leading towards the emergence of a man-like species. Man represents the quintessence of things primate; he possesses no physical characters that have not been anticipated by many of his predecessors. If he differs from his non-human relatives it is not in the basic nature of his characters, but in the degree to which they have been improved and refined.

I don't intend to give the impression that I think that destiny has been at work and that man's evolution has been purposeful. No-one who accepts the principle of the natural selection could accept the philosophy of orthogenesis, which was the last stronghold of the faint-hearted fundamentalists who, as scientists, sought a compromise between the belief in 'special creation' that had been their upbringing and the evolutionary principles of Darwin to which they were suddenly exposed. However, while rejecting metaphysical involvement, I do believe that there is a 'steering force' in evolution and it is a force which we call the environment.

The process might be called orthoselection. The environment operates as the principal agent of natural selection. Organisms within a lineage show a progressive improvement with time, in respect of their way of life, to which they become progressively more fully adapted.

The gibbons provide a clear example of orthoselection in action. The gibbons have evolved from a small, insectivorous, arboreal quadruped equally at home in trees or on the ground into a highly specialized arboreal ape which exploits the trees and their food resources to the full. Yet, in spite of this, gibbons have not lost 'the common touch'; they have not turned into a primate version of three-toed sloth condemned to hang upside down and provide a breeding ground for all the algae in the neighbourhood, neither are they koala bears or giant pandas, slaves to a specialized diet. Gibbons are not on the endangered species list; while there are trees there will always be gibbons.

The particular talent of the primates has been to avoid the traps of ortho-selection while enjoying its benefits. They have managed to become specialists without, thereby, sacrificing their freedom of action. They achieved this by *acquiring* some new general purpose characters, especially those related to the brain and to the special senses, and *retaining* the best of the old mammalian characters that they had inherited.

Man is a good primate. In spite of the fact that his distant Miocene ancestors changed their evolutionary direction by deserting the arboreal milieu for a terrestrial one. This transition resulted in only a single hiccup in the orderly progression from primitive mammal to man. What man's ancestors 'did' was to adapt a physical equipment—programmed for an arboreal career—to meet the exigencies of life on the ground. Paradoxically, most of man's arboreal attributes were ideally suited to life in a new environment. Only in the locomotor system were major changes necessary; his ancestors had to learn to support and propel themselves by their legs instead of their arms; this was the nature of the hiccup.

When did man emerge from the primates? The question is really irrelevant. He was there from the beginning.

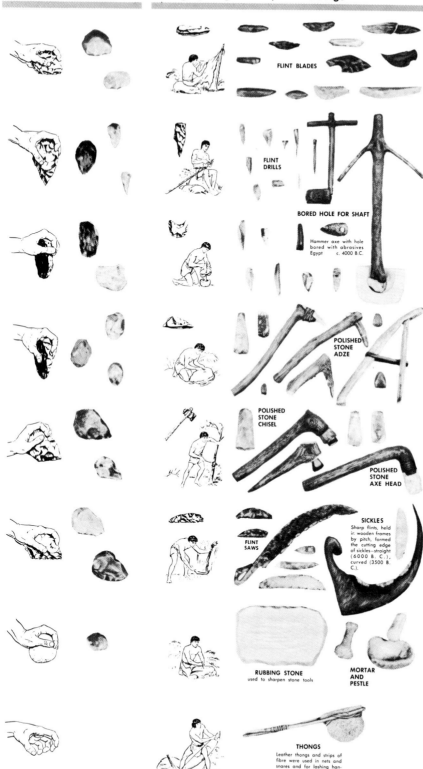

FLINT BLADES

FLINT DRILLS

BORED HOLE FOR SHAFT

Hammer axe with hole bored with abrasives Egypt c. 4000 B.C.

POLISHED STONE ADZE

POLISHED STONE CHISEL

POLISHED STONE AXE HEAD

SICKLES
Sharp flints, held in wooden frames by pitch, formed the cutting edge of sickles—straight (6000 B. C.), curved (3500 B. C.).

FLINT SAWS

RUBBING STONE
used to sharpen stone tools

MORTAR AND PESTLE

THONGS
Leather thongs and strips of fibre were used in nets and snares and for lashing handles to stone tools.

Ills. 96, 97. As early man lacked both claws and canines, he probably utilized natural objects—stones, bones and sticks—in order to defend himself (as do some primates today— see *Ill. 144*). Gradually he began to alter these objects to increase their usefulness. Initially he made crude, chipped pebble tools (opposite), then he learnt more refined ways of chipping and eventually started to use the flakes themselves. This sequence is shown in the left-hand column from bottom to top. As time went on he developed a functional tool-kit— chisels, drills, scrapers and saws. A real breakthrough occurred when he started to put components together, hafting axes or making a saw edge by mounting several small blades into a handle. The many varieties are shown in the right-hand column.

5 The First Revolution

Barbara Bender

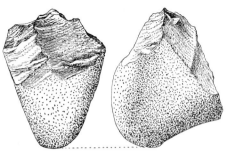

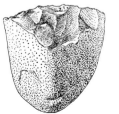

THE process of evolution never ceases. Man, like all other animals, adapts to the changing environment. But, unlike other animals, his adaptations are cultural as well as biological. The two processes not only function differently, they also proceed at different speeds. Biological change involves mutation and each mutation takes at least a generation. Cultural change needs an initial concept, the means to put it into practice, and social approval. Each innovation can be initiated quite rapidly.

Cultural change is a cumulative process. A primitive hominid with a restricted technology and a narrow range of tools has only a limited ability to conceptualize new tools. A man who possesses nothing but a pebble-axe is unlikely to conceive of an arrowhead. But with improved technology and a wider repertoire the potential for innovation increases.

It follows that as cultures become more complex, the pace of change quickens. However, this acceleration is not a smooth progression: both temporally and spatially it proceeds in jerks. Culture can be envisaged as an adaptive system made up of a number of interlocking and interacting parts, such as technology, economy, social organization and ideology. There will be long periods of time during which only minor shifts occur and the culture remains stable. But there will be other times when changes in one or more parts cause immense repercussions throughout the rest of the system.

This chapter attempts to show some aspects of the long process of cultural development. It begins with the slow opening of the Old Stone Age (Lower and Middle Palaeolithic), over three million years in which cultural development often lagged behind physical evolution. There follows an acceleration of cultural change associated with late hunters and gatherers from *c.* 20,000 years ago (Late Palaeolithic and Mesolithic). Then, with the development of farming (the Neolithic), *c.* 11,000 years ago, comes an incredible cultural burgeoning. Man increasingly modifies and controls his environment, and this control makes possible social organization on a scale and of a complexity far beyond the capacity of even the most advanced hunting-and-gathering society.

HUNTING-AND-GATHERING SOCIETIES

Hominid groups have existed in Africa for about four million years. The early groups had very limited cultural equipment—they did not differ greatly from other comparable species such as the apes. Their means of communication were undoubtedly rudimentary. Their tool-kit, roughly pointed bones and crudely chipped pebbles, had limited application for defence and as tools. They were unable to control fire. Teeth, nails and mobility were of para-

mount importance. The narrow cultural base precluded rapid development. Crude tools in the Omo valley in Ethiopia date to 4 million years ago, comparable tools from Lake Rudolf in Kenya date sometime before 2·6 million years ago, and hardly different tools from Bed I in the Olduvai Gorge in Tanzania date as late as 1·75 million years ago.

But gradually the hominids' brain and prowess increased. Communication improved. Knowledge and skills were handed down from generation to generation and the ability to conceive of and make tools increased. Cultural innovation gathered momentum.

Sometime around a million years ago hominid groups (probably early *Homo erectus*) spread out from Africa across most of the Old World. In many regions they faced climatic and environmental conditions very different from those of today, and often more extreme. For, sometime between one and two million years ago, the last great Ice Age, the Pleistocene, began and massive ice-sheets covered much of Northern Europe and North America. At their maximum extent they reached south to a line from Bristol to Berlin in Northern Europe and south of the Great Lakes in North America. Even areas far away from the ice sheets were affected. Within the Pleistocene epoch there were times when temperatures rose and the ice-sheets retreated, times when they fell again and the ice-sheets advanced once more. It was not until 12,000 years ago that they finally retreated and climatic conditions in Europe and North America gradually ameliorated.*

The hominid groups that moved out of Africa and across the Old World had to adapt their skills, equipment, economic strategy and no doubt their

Ill. 98. Early man not only learnt more refined ways of handling stone but he also improved his wooden artifacts. Unfortunately, due to the perishable nature of the material, archaeologists very rarely find wooden tools. One important innovation must have been the hardening of spear-tips by placing them in the fire and then scraping them. Here, a Western-desert Aborigine heats a spear-shaft in order to straighten and harden it. A similar object has been found in England.

* Some authorities would say that this was not a 'final' retreat but simply another relatively warm interglacial.

Ills. 99, 100. An early method of flaking was to strike the pebble against another stone held in the hand. It was difficult to control the process and the flaking was therefore crude. Pressure flaking was much more advanced. The tool was held against a piece of leather which softened the impact, and the flakes were removed by pressing rather than by striking. The diagram (below) shows a soft hammer of wood or bone being used, making it possible to obtain a delicate re-touching of the edge. The Acheulian hand-axe, shown life-size (right), would have been made by the first technique.

social organization to varied and often harsh environments. One vital adaptation that took place at least 500,000 years ago—or perhaps somewhat earlier—was the control of fire. Such control meant that not only was man able to cook his meat and break down the valuable fats and proteins, but he could also protect himself from wild animals, provide light, warm himself, and make effective weapons by hardening the points of wooden spears in a fire.

Ill. 198

Ill. 98

Already prior to 500,000 years ago, technology—both in terms of skill of manufacture and in range of tools—had begun to show very considerable advances. In the Acheulian culture, found mainly in Africa and Europe, cores began to be much more carefully chipped. Instead of striking a pebble against a rock set on the ground (the anvil technique), the pebble was chipped by means of another hand-held stone or piece of bone or wood. This allowed a much greater degree of control. Tools could be thinner, more carefully shaped and geared to specific functions. So hand-axes, cleavers, picks and chisels were now made. The flakes also were sometimes used.

Ills. 81–83, 100
Ill. 99

With such a tool-kit, it was possible to hunt a wider range of animals, including large game. At the site of Ambrona in Spain, dated *c.* 300,000 years ago, a herd of forty or fifty short-tusked elephants was driven— probably by setting the grass on fire—into a swamp.[7] There they foundered and stuck and were dispatched with fire-hardened wooden spears. Cleavers, hand-axes, bifacial tools and side-scrapers were used to carve up the meat and to smash open the skulls. Such a hunt would, one suspects, require not only technical adaptability and physical prowess, but also a reasonably developed system of communication and probably a degree of co-operation as well. Perhaps several small bands combined together for the event.

It may also be that the male/female roles became more clearly defined. With more organized hunting of larger game, women with young children would not be able to play much part. Moreover, for biological reasons (see p. 155) babies remained relatively helpless for longer than previously. The female was therefore less mobile and may have begun to concentrate on collecting wild plants, seeds, nuts and berries.

There may also have been developments on the ideological level. Human skulls from the great caves of Chou-k'ou-tien in China, dating *c.* 400,000 years ago, had artificially enlarged holes at the base. The cave inhabitants may have been cannibals, or it may be that some ritual was being observed— perhaps involving eating the brain of the dead to pass on strength, cunning, etc. Such rituals were certainly practised at a later date in South-east Asia.

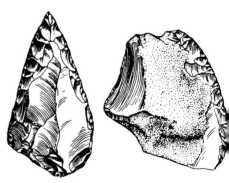

Ill. 101. A Mousterian point and side-scraper, showing the Middle Palaeolithic flake technique.

To give some idea of the time-scale of these developments, the five million years of hominid development, using the analogy of *Ill. 24*, can be represented as eighteen days. Changes so slow as to be almost imperceptible occupy the first fortnight. The Acheulian culture begins two days ago, and all the cultural developments from agriculture to skyscraper occupy the last half hour.

Towards the end of the Middle Palaeolithic there is increasing evidence of more effective hunting-and-gathering societies. In the Ancient Near East, *c.* 60,000 years ago, Mousterian groups lived mainly in caves, lit fires and had a varied tool-kit. They were able to conceptualize the sort of flake they required for a particular tool and could prepare a pebble prior to striking the flake. The flake was then retouched to achieve the final form. These Neander-

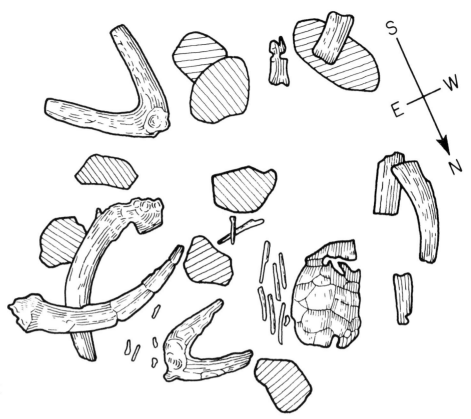

Ill. 102. The burial of a Neanderthal boy at Teshik-Tash, surrounded by goat frontlets.

thal men were proficient hunters, and butchering stations have been excavated where one or two animals were carved up after the kill. Other sites were base-camps with much more abundant and varied material, including many bones of various species, spreads of ashes where cooking was carried out, and concentrations of waste flakes where stone tools were manufactured.

In other regions, contemporary Mousterian groups lived out in the open. In Western Russia, for example, rough circles of hearths have been excavated, surrounded by rings of heavy elephant bones and tusks which probably served to weigh down skin tents.[3] There is also widespread evidence of purposeful burials. At La Ferrassie in South-west France two adults, a male and a female, were buried head to head. Close by two infants were placed in pits and a little further away there were small mounds, two of which contained children—one apparently a newborn babe.[7] In Southern Russia, at *Ill. 102* Teshik-Tash, Uzbekistan, a child was buried in a pit and surrounded by six pairs of horns from Siberian mountain goats.

By Late Palaeolithic and Mesolithic times, when the ice sheets covering Northern Europe and North America were beginning to retreat, man, though still dependent on hunting, fishing and gathering, was far removed from the early African hominid. Not only in time—nearly four million years had elapsed—nor in physical development—*Homo sapiens sapiens* had emerged—but also in cultural complexity and ability to adapt.

The cultural complexity permeates every sphere—economic, social and ideological. Different types of economic strategy can be distinguished. Some groups practised rather specialized economies, concentrating on a limited number of resources. For example, many French Magdalenian groups, living between 19,000 and 12,000 years ago, specialized in hunting herd animals, particularly reindeer, and some also concentrated on salmon fishing. Of course this specialization did not preclude the hunting of other animals and this is borne out not only by the bones found on the sites, but also by the cave art of Southern France and Northern Spain where the animals are painted or engraved with superb accuracy. This art also bears witness to the importance of ritual in the lives of these people. The positioning of the animals within the caves, the superimpositions, the choice of animals, the juxtaposition of symbols and animals, the painted outlines of hands, all suggest that, as in most primitive societies, the paintings had a significance far beyond a simple portrayal of nature.[22]

Other groups practised a 'wide-spectrum' economy in which numerous resources were exploited. Sometimes, though not always, such exploitation was associated with a complex equipment for procuring, processing and storing the wild foodstuffs. Sometimes there were a variety of projectile points for hunting different types of game, or sickle blades for cutting grasses, or querns and grinding stones for grinding nuts and seeds.

Many, indeed most, hunting-and-gathering groups were mobile. Sometimes, as in the case of the French groups, they had to follow the migrating herd animals; sometimes the wild resources were scattered. Such mobility was not a matter of aimless wandering. There is good evidence that many

Ill. 103

Ills. 6, 38, 104

Ills. 103, 104. The reconstruction of an Upper Palaeolithic family group brings together an unlikely mixing of Cro-Magnon hunters, a steatopygous Predmost woman and an elderly Chancelade man (below). Rock shelters such as the one depicted were used by Magdalenian hunters, as were the deeper caves. In such rock shelters and caves Magdalenian artists depicted a wide variety of game, as well as symbols and a few human figures (opposite, and see *Ills. 6, 38*).

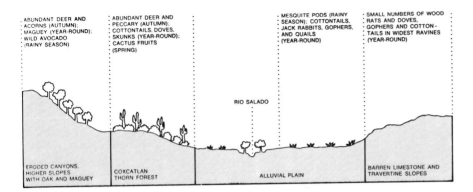

ABUNDANT DEER AND
ACORNS (AUTUMN);
MAGUEY (YEAR-ROUND);
WILD AVOCADO
(RAINY SEASON)

ABUNDANT DEER AND
PECCARY (AUTUMN);
COTTONTAILS, DOVES,
SKUNKS (YEAR-ROUND);
CACTUS FRUITS
(SPRING)

MESQUITE PODS (RAINY
SEASON); COTTONTAILS,
JACK RABBITS, GOPHERS,
AND QUAILS
(YEAR-ROUND)

SMALL NUMBERS OF WOOD
RATS AND DOVES,
GOPHERS AND COTTON-
TAILS IN WIDEST RAVINES
(YEAR-ROUND)

RIO SALADO

ERODED CANYONS,
HIGHER SLOPES
WITH OAK AND MAGUEY

COXCATLAN
THORN FOREST

ALLUVIAL PLAIN

BARREN LIMESTONE AND
TRAVERTINE SLOPES

late hunting-and-gathering groups had a carefully scheduled round geared to the exploitation of plants ripening at different times in different places. One such annual round has been pieced together in the Tehuácan valley of Mexico.[12] Richard MacNeish and his co-workers divided up the valley into its component micro-environments and then analysed the remains from sites in these environments in order to establish at what time of year they were occupied and what resources were utilized. They found that the El Riego hunters and gatherers who lived in the valley *c.* 8500 years ago used to spend the dry winter months in the valley bottoms and on the alluvial slopes, where they concentrated on hunting and trapping, supplemented by cutting opuntia and agave leaves. Then in the spring they scattered more widely, searching for seeds and pods. In summer, when conditions were wetter and resources more abundant, they congregated in the valley bottoms and collected seeds. Finally, in the autumn they moved up the alluvial slopes, mainly gathering fruit but also hunting and cutting leaves. It is possible that some groups had begun to cultivate a few crops—perhaps a few avocados, chilli peppers, amaranths and squash—but, if so, this had no effect on the pattern of seasonal movement.

The analysis of the Tehuácan sites also throws some light on the social organization. In the lean months the El Riego groups moved around in small units—perhaps just a couple of families—but in the summer, when resources were more abundant, they amalgamated into bigger units.

Ills. 105, 106

Ills. 105, 106. The Tehuácan valley, in Central Mexico, has been studied in great detail. Above left: a simplified cross-section of the valley shows the different micro-environments and their seasonal resources. Below is a reconstruction of the seasonal movements of the El Riego hunting-and-gathering groups *c.* 8000 BC. In the leaner seasons, indicated by the small quartered circles, the groups split into small units. Then, in the summer, when in favoured areas resources were more abundant, they came together at larger camp sites.

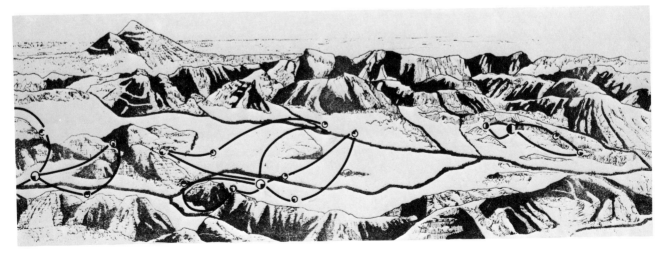

Not all late hunters and gatherers were mobile. Contrary to popular belief, village life began before farming developed. Hunters and gatherers living in 'optimal' areas, where wild food resources were both abundant and concentrated in a limited area, were able to settle in one place all the year round. *Ill. 107* For example, there is the village of Lepenski Vir in Yugoslavia, dramatically sited on a terrace within the rugged Iron Gate gorges of the river Danube.[20] The inhabitants seem to have lived almost entirely on fish. They grew no crops and kept no animals. They apparently speared the fish with flint-tipped shafts or stunned them with large stone clubs. Mammoth sturgeon and carp, sometimes several feet long, are still found in the Danube and it is not surprising that, with such resources, the Stone Age inhabitants had no need to move around. They built large trapezoidal houses with well-paved floors and stone-lined hearths. It was beside some of the hearths that the great sculpted boulders were found, some with faces, others with what seem like abstract designs. Some may represent a fish-like deity. As with the cave art in France and Spain, there is probably some connection here between the ritual representations and important food resources.

Ill. 107. A close-up of one of the hearths at Lepenski Vir, with the huge sculpted boulders alongside it.

Ills. 108, 109. At Munhata, in the Jordan valley, animals were domesticated but there is no evidence of cultivated crops. Despite this, the equipment for grinding the wild cereals (left, a stone quern, and right, a stone mortar) is quite elaborate.

Ain Mallaha is another hunting-and-gathering village, this time in the Upper Jordan valley.[15] The settlement was close to Lake Huleh, on the edge of a vegetation zone where oak and pistachio grew and where wild barley and wheat flourished. Such wild cereals sometimes extend over thousands of hectares and give high yields, between 500 and 800 kilograms per hectare in a good year. An American archaeologist, Jack Harlan, tried harvesting wild wheat with a stone-bladed sickle and found that, although only an amateur, he was able to collect a kilo of clean grain in an hour.*[5] This means that a family of experienced plant collectors could collect a metric ton during the three weeks that the grain was ripe, enough to last them right through the year. At this site no plant remains were preserved, but there were mortars and pestles, blades with 'sickle-gloss' (a polish that results when grasses or reeds are cut), and plaster-lined pits which may have been used for storing grain. The inhabitants probably gathered the wild cereals, and they certainly fished in the lake and hunted, particularly gazelle. These wild resources were sufficient to support fifty households, perhaps two or three hundred people.

* The experiment was actually carried out in South-east Turkey but the results are just as applicable to the Jordan valley.

At this site there is very little in the way of art, but on the other hand there is a hint of a more complex social structure. There is one particularly well-made hut that might have belonged to a 'head-man'. There is also one rather elaborate burial where a male and probably a female (at least 'she' had a necklace) were placed in a pit grave which was covered by a pavement surrounded by stones. On the pavement a fire was lit, then earth was heaped over and another platform constructed.

It is possible that, in rare instances, a hunting-and-gathering economy was able to sustain a settlement larger than a village. The evidence, though meagre, is worth considering. On the plain of Konya in Anatolia there is a large 'tell' called Çatal Hüyük. An important excavation was begun there by James Mellaart,[14] but it was not completed and the bottom of the tell remains untouched. In the lowest excavated level (level X), which dates to *c.* 6000 BC, part of a flourishing township was uncovered. The inhabitants were *not* hunters and gatherers. They grew many different crops and kept herds, including large cattle. The town's prosperity, however, was probably not due to the agricultural base but to its position as a great trade centre. Obsidian was available close at hand, and from this black volcanic glass, which is very easy to chip, the inhabitants made beautiful weapons and tools. They also used greenstone, alabaster, and lava, all available in the vicinity. They obtained marble from the West, flint from Northern Syria, large cowries from the Red Sea and copper from the Tauros mountains. The township was rich enough to support a priesthood and there is much evidence of an important bull cult.

Ills. 110, 111

Ill. 113

Ills. 112, 123

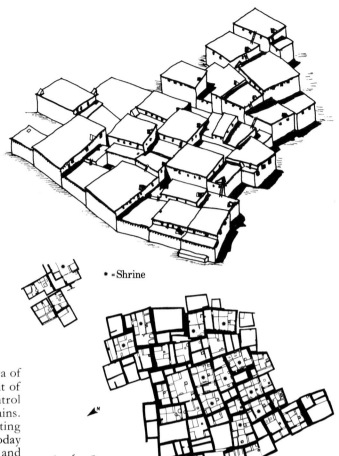

* = Shrine

Ills.110–112. Right: plan and reconstruction of a small area of the township of Çatal Hüyük (level VI). The development of this highly complex town may well be linked with the control of the obsidian found in the neighbouring volcanic mountains. From evidence of a bull cult (below) and of murals depicting hunting scenes, it seems probable that the area, which is today mainly arid and saline, once supported abundant game and plant life.

All this may seem to be irrelevant in the context of hunting-and-gathering settlements. But the question is, what sort of economy did the earlier inhabitants of the tell practise? Below the excavated strata lie many metres of unexcavated occupation. The initial occupation must have begun very much earlier, perhaps several millennia. It probably pre-dates any evidence of domesticated plants and animals in the rest of the Ancient Near East. Is it possible, therefore, that the settlement belonged to a hunting-and-gathering community? Could it already have been a trade centre for obsidian? Could sufficient wild food have been brought in as barter to support a settlement large enough to be called a township?

There is another large tell not far away, Asilki Hüyük, which has earlier Carbon 14 dates than Çatal Hüyük. They range between 6600 and 7000 BC. There is no pottery. It has not yet been excavated but there is much obsidian on the site. Was it another trade centre? What was its subsistence base?

Further to the South there are many great tells situated at nodal trading points, such as Nineveh and Kirkuk. Their basal levels have never been excavated.

Obviously we need more information in order to follow up these clues. These sites are mentioned only to suggest that it is possible—though not proven—that a hunting-and-gathering economy could support large sedentary communities.

This account of the variety of late hunting-and-gathering societies should have dispelled any notion that such people were necessarily miserable, half-starved or forever on the move. Rather, they were highly adapted and often very successful societies. Cultural change could be initiated rapidly, unlike earlier hunting-and-gathering cultures which remained virtually unchanged for hundreds of thousands of years. Compare, for example, the early hunting-and-gathering Acheulian culture which lingered on for *c.* 425,000 years (500,000–75,000 BP) with the late hunting-and-gathering Zarzian culture of the Ancient Near East, which lasted no more than a thousand years (12–11,000 BP).

But even the most complex late hunting-and-gathering societies were dependent on wild resources. They were circumscribed by what was naturally available. Permanent settlement was only possible where food resources were concentrated and abundant, or perhaps where there were tradeable commodities that could be exchanged for foodstuffs. Population densities were limited by what was available during the leanest seasons, although to some extent this may have been offset by storage.

To counteract these limitations, man had to produce his food. Then features that were exceptional in hunting-and-gathering societies and only possible in 'optimal' areas could become the rule and complex societies unknown in hunting-and-gathering contexts could develop.

Ill. 113. A fine, pressure-flaked flint dagger, with a bone handle carved in the shape of a snake. Not only was Çatal Hüyük a great trade centre, it was also an important manufacturing centre and produced a great range of highly sophisticated objects.

EARLY FARMING SOCIETIES

The archaeological evidence of food production goes back no more than 11,000 years. In that short time-span the rate of cultural change has been vertiginous. Before 11,000 years ago all men were hunters and gatherers, totally dependent on wild resources. Since then food-producing economies—based on crop cultivation and/or animal herding—have spread across the length and breadth of the world. They have transformed the relationship between man and the land and man and other animals. Villages have become the primary form of settlement. In a mere 400 generations, all the great civilizations, all the great religions, urbanization, and industrialization have emerged.

The transformation from hunting and gathering to food production was not an abrupt phenomenon. It was part of a long process. The domestication of plants and animals is one among many types of plant and animal manipulation. It evolved out of earlier hunting-and-gathering strategies. There were undoubtedly hunters who killed herd animals selectively and thus practised a form of stock-culling. There were probably gatherers who tended the plants without actually cultivating them. Until quite recently, the Paiute Indians living in the Owens valley of the South-west U.S.A. dug ditches and dammed streams in order to improve the yields of their wild crops.[21] They never sowed seeds. It is a small move from such manipulative practices to more deliberate herding and breeding, to the planting out of a few plants on fertile river banks (horticulture) or the strewing of a few seeds on small well-watered patches of ground (agriculture). But unlike the other practices, these resulted, through the effects of selection and/or captivity, in genetic changes. New races and species emerged. And so domestication begins.

There will never be a simple answer as to why some groups made the move to food production. The explanations will vary from area to area and from group to group. The move to herding would have different antecedents from the move to cultivation. The move to seed-crop cultivation is unlikely to involve the same processes as the move to root-crop cultivation. Even the move to seed-crop cultivation in two regions with fairly similar climates (the

Ills. 115, 120 Near East and Meso-America) would not follow the same course. What would be grown or herded would depend on what was available, on the knowledge and technical ability of the groups involved, and on prior hunting-and-gathering strategies and biases.

Whatever the reasons, it is becoming increasingly clear that there was not one centre of early food production, nor two, but rather a great many. The shift from hunting and gathering to food production was repeated independently over and over again. There is archaeological evidence from places as far apart as the Near East, Meso-America, upland Peru and interior Thailand.[2]

The shift was gradual and the importance of food production would not have been immediately apparent to the early practitioners. Early attempts at cultivation must have been full of uncertainties and set-backs. For generations, crops and herd animals would have had to be supplemented by hunting and gathering. Indeed in Meso-America, where there were virtually no animals suitable for domestication, even such advanced societies as the Maya and the Aztecs continued to rely on wild game for their meat supply.

But with hindsight we can acknowledge the significance of domestication and the rest of this chapter will be devoted to unravelling some of the more important implications.

Carrying Capacity. Once man began to grow crops or to herd animals he could transfer them beyond their natural habitats and extend their environmental range. He could concentrate a variety of plants and animals within a limited area. He was therefore in a position to create artificially or extend 'optimal' areas. Then he could remove some of the pressures of natural selection so that deviant mutations could survive, and eventually he could learn to select beneficial mutations. For example, wild barley (and most other *Ill. 93* wild winter cereals) has a brittle rachis (axis) between ear and stem so that in the wild the ear easily falls off and the seeds are dispersed. This makes it very difficult to harvest since the ears are liable to fall off when the stems are touched.[17] But a man harvesting wild barley with a stone sickle unwittingly collects those with tougher rachis because they stay intact. Without any knowledge of selective breeding, he selects a mutant disadvantaged in nature, but one that allows him to harvest more effectively. The wild barley grain is also difficult to thresh because it has a tightly fitting husk. A simple mutation frees the grain from the husk and gives rise to 'naked' varieties. These, too, would tend to be selected and a deviant mutation thereby favoured.

The case of maize is still more dramatic.[19] Early cultivated maize had *Ill. 114* miniscule cobs, no more than 20 mm. long (less than two postage stamps!).

Ill. 114. This picture graphically shows the dramatic increase in size from the primitive maize cob to the modern-day hybrid.

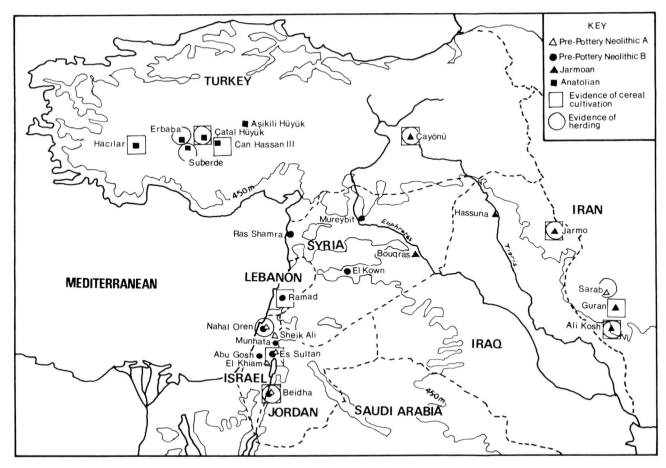

KEY

△ Pre-Pottery Neolithic A
● Pre-Pottery Neolithic B
▲ Jarmoan
■ Anatolian
□ Evidence of cereal cultivation
○ Evidence of herding

Ill. 115. Map showing the early farming sites in the Ancient Near East.

There were only eight rows of tiny kernels, each partially enclosed in a glume; the rachis was brittle and each cob was partly enclosed in a few husks. But maize has remarkable genetic flexibility and can cross with other wild grasses. Gradually, through crossing and back-crossing, domesticated maize increased enormously in size, the number of kernels shot up, the rachis became larger and tougher and the glumes were reduced. At the same time, the whole ear became encased in overlapping husks. The final result, a modern cob, is ten times the size of the early Mexican cob, but because of the enclosing husks, is totally dependent on man for seed dispersal.

All these factors—transplantation, concentration of plants and/or animals and selection of mutations beneficial to man—meant that the land could be made to support a larger population. Obviously, in areas which had been optimal for hunting and gathering, the carrying capacity of the land was already high and early food production would have had little effect. The importance of food production is that it *extended* the areas of high carrying capacity.

The above reasoning would suggest that the shift to food production stimulated population growth. For if the carrying capacity is increased it follows that the population can increase. This sequence is questioned by some authorities. E. Boserup makes a strong case for reversing the order.[4] She maintains that it is population pressure which triggers off economic innovation. Take for example a hunting-and-gathering society living in an optimal resource area. If they settle down the population will probably increase (see pp. 123, 124) and a time will come when there is real pressure on the food resources. They will then have three alternatives: they can emigrate, they can inaugurate population controls, or they can innovate in order to get greater returns from the land. A possible innovation would be to produce food. If this occurs it could be claimed that population pressure has led to economic innovation. A similar case could be made where environmental change depletes the food resources and results in population pressure. The question of which comes first, the pressure or the innovation, would seem like the puzzle of the chicken and the egg. Population pressure stimulates innovation: innovation stimulates population pressure.

Of course the carrying capacity would only increase gradually. In the Ancient Near East, for example, the archaeological evidence suggests that *Ill. 115* cultivation and herding began not on the fertile alluvial soils of the Tigris/ Euphrates where the later Mesopotamian civilizations arose, but in the foot-hills of the Zagros and Tauros mountains, the hills of the Levant and the intermontane basins of the Anatolian plateau. In these areas, rainfall was low and uncertain and early cultivation using primitive techniques was limited to areas with reasonable water supplies. At the early farming site of Ali Kosh *Ills. 116, 117* in Iran, the villagers grew crops so close to the marsh that quantities of club rushes were included in the harvest.[6] On the other side of the world, in Meso-America, the evidence of early cultivation comes from the high arid basins and valleys of the interior Mesa and from the flanking slopes. These are rain-shadow areas and *Ill. 106* shows the sort of inhospitable terrain *Ill. 106* involved. Here, early cultivation seems to have been concentrated along the river banks. Only when crops and techniques began to improve, when simple irrigation or drainage was practised or traction animals were used, could more land be brought under cultivation. At Choga Mami, in Iraq, irrigation was under way by 5500–5000 BC, only 1500 years after the earliest evidence of cultivation in the Near East. With irrigation, farming could spread down on to the piedmont that flanks the Zagros and Tauros to the south, and shortly afterwards, when irrigation was combined with drainage, the alluvial soils of the Tigris/Euphrates valleys could also be utilized. It has been estimated that the average Late Pleistocene hunter-and-gatherer carry-ing capacity in the Ancient Near East was probably no more than 0.1 person per square kilometre. This increased to 1 or 2 persons under early dry farming and with the development of irrigation it increased six-fold. If we then add the effects of later innovations in food production emanating from cities and large-scale industries, the increase in world population becomes quite staggering. A world populated by hunters and gatherers could probably not have contained more than 20 to 30 million people. The modern world contains over 3,500 million.

Permanent Settlements. The increase in the carrying capacity goes hand-in-hand with an increase in permanent settlement. We have already seen that hunters and gatherers could become sedentary in 'optimal' areas where wild resources were abundant and concentrated. In nature such areas are excep-tional. Food production mimics these optimal conditions, extends the area of high carrying capacity and thereby allows the exception to become the rule.

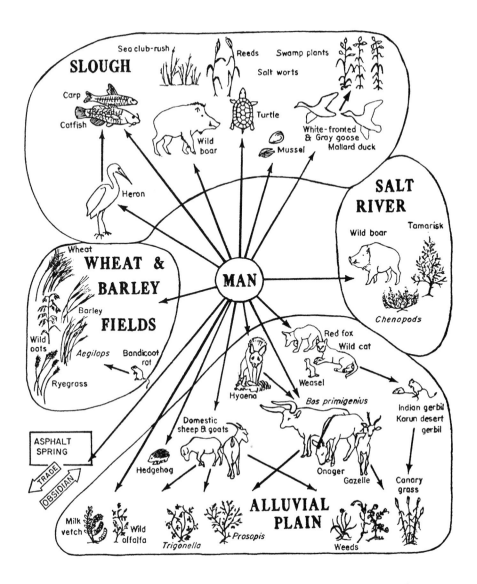

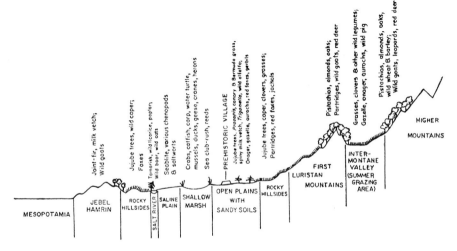

Ills. 116, 117. Ali Kosh was a small, early farming village in a fairly unfavourable environment. The cross-section (below right) shows the micro-environments with their resources, and the utilization of the latter is more graphically represented above.

Ills. 118, 119. With the development of sedentary life, associated with farming, house structures often became more permanent and more complex (see *Ills. 110, 111*). In some cases, as the community grew larger, defensive works and irrigation had to be undertaken, and there was an increasing need for a more corporate social organization. Çayonu (left) was a small village but had surprisingly elaborate architecture, while Jericho was a large, strongly fortified township, as can be seen from this massive round tower, which had an internal staircase (above left).

Ill. 115

Ill. 119

Ill. 118

Ill. 120

Again this is a gradual process: the pace depends on local conditions. In the Ancient Near East occasional permanent settlements were already established by hunters and gatherers in the ninth millennium BC. With food production based on a combination of high-yielding winter cereals and herding, the number increases very rapidly. By the eighth and seventh millennia BC there were settlements scattered through the Levant, Anatolia, and the Zagros and Tauros hillslopes. Some were already sited in marginal areas beyond the natural habitat of the wild cereals (although of course the modern distributions may not be quite the same as those of the seventh millennium). Groups in less favoured areas may have turned to food production in an attempt to improve upon the inadequate wild resources. These settlements were mainly small, with probably no more than 200 inhabitants. But already in the eighth millennium BC there was one township which covered 4 hectares, probably housed 2,000 to 3,000 people and was defended by a massive wall, a tower and a flanking ditch.[8] This was Jericho and it is an enigma. Why should a community, living beside an oasis in a desolate, arid region need such defences? What did they live on? For although there is evidence of domesticated wheat, the area that could be cultivated by dry farming was limited to a tiny patch around the oasis. Could they already be irrigating? Was it a trade centre—like Çatal Hüyük a millennium later—and did cultivated grain come in as barter? It is strategically placed between the Judaean hills and the semi-desert, with good connections to the north via the Jordan valley. Some fine imported stone—obsidian, nephrite and greenstone—has been found on the site but only in small quantities. Perhaps the trade was in perishable commodities such as salt, bitumen and sulphur.

This rapid establishment of permanent settlements, both large and small, in the Ancient Near East stands in sharp contrast to developments in highland Meso-America. Here the wild resources were scattered and not particularly high yielding. There were no sedentary hunters and gatherers. Early food production did little to alter this pattern; there were no animals suitable for herding and the early cultigens, including maize, were low yielding. Wild resources therefore remained important and seasonal mobility remained essential. Early food producers, in the sixth millennium BC, grew no more than five per cent of their food, and had no permanent settlements. A thousand years later, there were still no settlements and only fourteen per cent of their food was grown. Two thousand years later, sometime between 3400 and 2300 BC, a quarter of the food was grown and the first slightly more permanent settlements appeared—perhaps still only occupied for part of the year. Only *c.* 1500 BC was the maize yield sufficiently high and the combination of maize-squash-beans sufficiently successful for the people to be able to settle down in small hamlets.[13]

Population Growth and Social Organization. Permanence of settlement has a direct bearing on population increase. In nomadic societies there is a premium on mobility and births are generally spaced at three- to five-year intervals.* When people stop moving around, this wide spacing is no longer necessary and the birth rate may rapidly increase. Food production may bolster this trend by providing more assured and perhaps more nutritious food, so that infant mortality is reduced.

* Richard Lee shows that among the !Kung Bushmen of Botswana the females carry their children until they are four years old.[11] They would therefore find it very difficult to cope with more than one child under that age. They do not wean the children until they are three or four years and the long lactations act as a form of birth control. There may also be infanticide.

The land yields more, the population becomes increasingly stable, and not only the over-all population density but also the size of population aggregate increases. Larger groups require more 'policing' than smaller ones. Land clearance requires a co-operative effort. So does the defence of crops and herds. Formalized co-operation becomes more and more essential. So while hunting-and-gathering societies tend to be structured on a rather flexible kinship system and have few mechanisms for controlling strife, agricultural societies tend to be organized on a more corporate basis. In the Near East, by the sixth millennium BC, there were sites (Yarim Tepe and Umm Dabaghiyah, both in Northern Iraq) with large, well-planned complexes of rooms which probably served as communal storage rooms. We may suspect that such corporate efforts would quickly generate leaders and at least a minimal hierarchy would be established. When more elaborate techniques begin to be used, particularly irrigation and drainage, the need for over-all organization would become still more important.

Trade. A corollary of even modest social stratification is an increase in trade. The more important members of the community will wish to emphasize their position by obtaining 'luxury' goods such as fine stones, pigments or shells. In the Ancient Near East the percentage of obsidian in circulation increased very rapidly quite shortly after farming began. Any increase in trade would of course greatly benefit communities that control and exploit the raw materials and also any individuals acting as exchange agents.

Ill. 121. Opposite: unlike Jericho, most early settlements would not have had massive, stone-built defences, but would have relied on thorn and brush fences to keep domesticated animals in and wild animals out. However, the evidence for this is largely conjectural because of the impermanence of the material.

Ill. 120. Map showing the main centres of early agriculture in Meso-America.

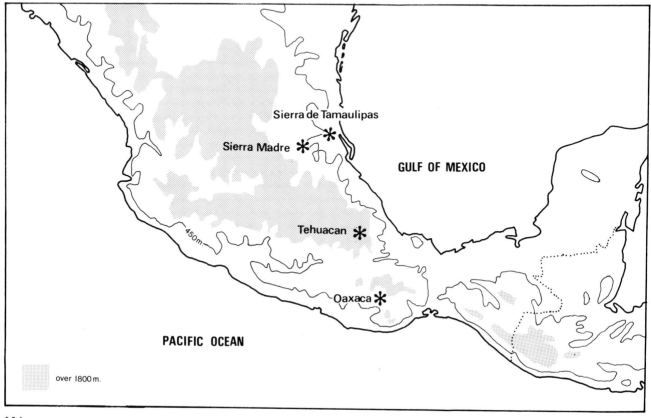

Crafts. There is also a connection between food production, the development of new crafts, and an increasing emphasis on craft specialization. In some instances the important factor is the increased permanence of settlement—because pottery is very fragile and weaving looms are bulky they are unlikely to be developed by nomadic people. Settled hunters and gatherers could, and did, make pottery. For example, there were hunting-and-fishing groups in Japan who made pottery several thousand years before agriculture was introduced. Conversely, mobile food producers did not make pottery. In the Tehuácan valley of Mexico, cultivation incorporated in a semi-nomadic existence began *c.* 5000 BC; permanent settlement probably began *c.* 3000 BC; pottery began to be made *c.* 2300 BC.

Ills. 105, 106

In many instances trade acts as a stimulus to craft innovation and specialization. Manufactured goods will be very useful for bartering against fine stones, shells, etc. In the Ancient Near East we have already seen how trade in obsidian increased shortly after the inception of food production. At the same time craft 'shops' began to appear. As early as the eighth millennium BC[9] Beidha in Jordan may have been a small regional bazaar. This little settlement comprised one or more large houses with finely plastered and painted walls separated by courtyards from a huddle of tightly packed workshop units. Each of these units had a long corridor and a series of cubicles, and many of the latter served specialized functions—horn- or bone-working, bead manufacture, even a butcher's shop!

Ill. 122

In many accounts of early farming it is suggested that food production brought about 'increased leisure'. This is hard to prove. In many instances the opposite is probably true—food production was more labour intensive and so reduced the amount of 'leisure'. A recent study has shown that the !Kung Bushmen, living in marginal conditions in Botswana, spend only twelve to nineteen hours a week searching for wild plants and animals, which is a great deal less time than the average primitive farmer (or the average

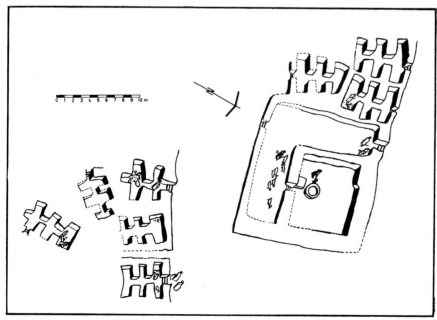

Ill. 122. A partial reconstruction of the regional bazaar at Beidha.

Ill. 123. The aurochs seems to dominate the mythology of the people of Çatal Hüyük. Although a very large beast, it is shown out of all proportion to the hunters milling around and to the deer at the bottom of the painting. Dogs are also represented, although they are hard to recognize, and they may have been domesticated for hunting or herding.

Ill. 123

Ill. 124

'modern' man![10]). At most, the advantage of farming is that it created longer 'blocks' of free time between seasonal activities such as planting, weeding and harvesting. Early farming communities would not have had full-time craftsmen. Pottery-making would probably have been undertaken by the women, while other crafts were probably practised in the 'free time' left over from agricultural pursuits.

Ritual. With the increasingly complex social organization often goes a more formalized approach to ritual and religion. With permanent settlement comes permanent ritual buildings. Already at Beidha, in the eighth millennium BC, there were curvilinear buildings a little way beyond the settlement that served a ritual purpose.[9] Somewhat later a whole quarter of the township of Çatal Hüyük was given over to a bull cult.[14] Walls had paintings and reliefs, bucrania were stuck on the walls or on platforms. At Jericho, at much the same time, there was an elaborate cult of the dead.[8] Skulls were given modelled features, and the eyes were inset with shells.

We have now covered most of the phenomena directly or indirectly associated with early food production. We have seen how the shift to cultivation and herding meant that the land would support more people, and that there could be more permanent settlement, larger population aggregates, more complex social and religious institutions, increased trade and new crafts. All these cultural aspects interlock, none stands alone. And although it has been stressed that these developments did not occur overnight, still, in contrast with earlier cultural developments, they occurred comparatively quickly—within a few millennia—and they set the scene for the inception of 'civilizations'.

It is beyond our brief to examine the later ramifications of food production. Until quite recently a village farming economy was the main form of subsistence throughout the world. Only in a few areas were there more complex systems requiring market centres, often, though not always, in urban settings.

A 'complex system' is a less loaded term than 'civilization'. But neither is easy to define. One authority stresses that civilization is 'the crystallization of executive power',[23] another that 'civilized society is above all stratified

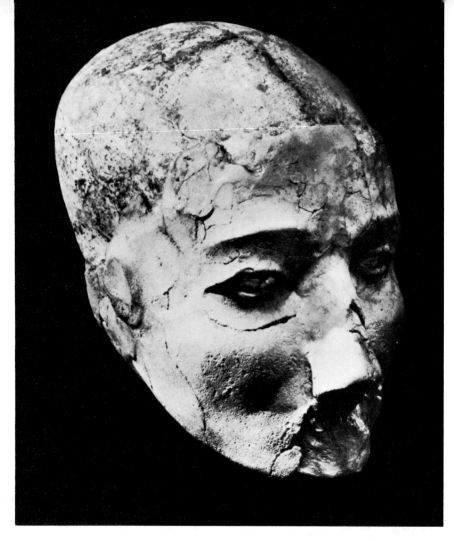

Ill. 124. One of the finely modelled skulls from Jericho—the eyes are inlaid with cowrie shells imported from the Red Sea.

society',[18] a third—more metaphysically inclined—defines it as 'the complex artificial environment of man, it is the insulation created by man, an artefact which mediates between himself and the world of nature'.[16] It is agreed that writing is not essential (the Incas of Peru could not write), nor is urbanization (the Maya had no towns). There is generally a high degree of technological ability and there is usually monumental architecture.

Whatever the precise definition, one thing is clear: no hunting-and-gathering group ever achieved 'civilization'.* And whatever the complex causes behind the rise of the different civilizations, whether it be the effect of population pressure or intensive farming, water control or trade monopoly, all have one feature in common: 'Complex, civilized societies depend upon a subsistence base that is sufficiently intensive and reliable to permit nucleated settlements, a circumstance that . . . in the long run has implied agriculture'.[1]

To return to our earlier theme. Only 400 generations of human beings, a tiny fraction on the time-scale of hominid development, have been food producers. But man, by gaining control of his food resources, has altered, adapted and complicated every aspect of his cultural make-up on a scale and at a speed far beyond those of physical evolution.

*The nearest were probably the North-west-coast Indians who had chiefdoms and complex rank systems.

Patterns of Behaviour

6 The Chimpanzee

*Jane van Lawick-
Goodall*

THE patient and dedicated work of the palaeontologists has, over the years, revealed a wealth of information concerning the structural evolution of man. The fossil evidence enables us to reconstruct the physical characteristics of our remote ancestors, to speculate on the use of the various tools and other artifacts found at their living floors with at least some items of their diet, and even to make informed guesses as to the size of their groups at different times. But palaeontology cannot tell us how Stone Age man looked after his children, how he interacted with his fellows, how he treated his old mother.

In recent years some scientists, in a new attempt to fill in the gaps in our understanding of early man and his behaviour, have turned to the study of man's closest living relatives, the non-human primates. Information as to the behaviour of different primate groups and their adaptation to differing environments helps us to build up, bit by bit, a more complete knowledge of the evolution of primate behaviour and is helpful to those concerned with trying to interpret the various strands of the palaeontological evidence. Baboons, because they are large aggressive monkeys living in highly organized social groups and often occupying the kind of savannah habitat thought to have nurtured emerging man, have frequently been studied by anthropologists interested in prehistory.[6] However, whilst we can learn much from observing how an intelligent primate copes with the dangers of such an environment, baboons are not nearly as closely related to man as are the great apes—the orang-utan, the gorilla and the chimpanzee. Of these three it is the chimpanzee, both in structure and behaviour, who is the closest of all.

Research in a variety of biochemical and medical fields has revealed striking similarities between men and chimpanzees, such as the number and the form of chromosomes, the proteins of the blood and the structure of the genetic material (DNA) itself. Chimpanzees are susceptible to the same contagious diseases as man. Perhaps most important of all, the anatomy of the chimpanzee brain is more like that of the human brain than is that of any other living creature.

What of his behaviour? Fourteen years ago the performances of laboratory and zoo chimpanzees, as well as those of a few brought up in human homes, had already indicated close behavioural similarities between this ape and man. Virtually nothing, however, was known about his way of life in his native homeland, the equatorial forest belt of West and Central Africa.[27]

Dr. Louis Leakey was among the first to realize the potential of the chimpanzee as a suitable model for the better understanding of the way of life of early man. It was he who sent me, in 1960, to try to observe the

Ill. 125. At some stage in hominid prehistory man's ape-like ancestors began to travel in the uniquely human upright posture. We can only speculate as to why this happened, just as we can only speculate on other distinctive patterns in human evolution. It is of interest, however, that the chimpanzee is capable of developing an extraordinarily upright striding gait. This is particularly true for Faben, shown opposite, who learnt to travel upright for long distances after losing the use of an arm during a polio epidemic. The bipedal posture prevents his arm from trailing along the ground.

chimpanzees of the Gombe National Park (then a game reserve) in what is now Tanzania. It took months for the timid chimpanzees to accept my presence, but eventually I was able to start learning details of their way of life and their social behaviour.

Over the years the research at Gombe has demonstrated that, in many aspects of his behaviour in the wild, the chimpanzee does indeed show remarkable similarities to human behaviour. He has a lengthy childhood, he forms close and enduring attachments within his family, some of his gestures and postures are almost identical with some of those shown by man, and are used in similar contexts. Moreover, he uses and makes primitive tools and he hunts in a co-operative group.

Whilst we cannot, of course, rule out the possibility of parallel evolution in man and chimpanzee, the total array of similarities between the two species, both physiological and behavioural, suggests that at some time, in the remote past, man and chimpanzee diverged from a common ancestor. If this was, in fact, the case we can postulate that those characteristics shared by modern chimpanzee and modern man were also present in that common ancestor and, therefore, in Stone Age man himself. We cannot assume that early humans behaved like modern chimpanzees, but we can hypothesize that in those far-off times, before the evolution of a sophisticated language and increased reliance on technology and culture, the behaviour of our human ancestors may have been remarkably similar to that of chimpanzees today.[21]

An understanding of chimpanzee behaviour is not only useful with reference to the behaviour of our ancestors; in addition it helps to highlight certain aspects which are unique to man and which differentiate him from the other living primates. Once we have realized these essentially human characteristics, we can then look for precursors in the behaviour of the chimpanzees and, if we find them, ask what evolutionary and social pressures

Ill. 126. The only unit in the chimpanzee community that is stable over a period of years is a mother and her dependent young. Here, two mothers have formed a temporary association. Evered (centre), a late-adolescent male of about twelve years, accompanies his mother, Olly, (left) and infant sister, Gilka. Flo (right) is seen with her dependent daughter Fifi (behind her) and small infant Flint.

Ill. 127. A chimpanzee constructs a nest, or sleeping platform, in the trees for the night. This takes him about three to five minutes and usually he makes a new one each night. Youngsters up to about five years of age share night-nests with their mothers.

Ill. 126

might have helped to shape such traits, through natural selection, towards their uniquely human form.[21]

Before describing some of the chimpanzee behaviour which is so like our own, let me briefly outline the social structure and way of life of the chimpanzees at Gombe.[18,19] The population, probably about 150 or so in the 30 square miles of rugged mountainous country running along the shore of Lake Tanganyika, is divided into communities of individuals. Within a community of some 30–40 members the chimpanzees recognize each other, probably by sight and by voice as well. The home range of a community, over a year, is about 10 square miles or possibly a little larger, and within this area the chimpanzees move about in small temporary associations, membership of which is continually changing as individuals split off to join other associations. These groupings may comprise males only, or females and youngsters, or combinations of both. There is no stable pair bonding between a male and a female—a mother with her dependent offspring (up to eight years old) forms the only association that may remain constant over a period of years. A family unit of this sort frequently travels on its own, but joins up from time to time with other associations. Some chimpanzees in a community meet but seldom, when circumstances such as a local abundance of food or a sexually attractive female draw them together; others meet more often; some show strong bonds of mutual attraction and associate frequently, particularly members of the same family.

Chimpanzees are omnivorous, but feed mainly on a variety of plant material, especially fruits. They also eat insects, honey, occasional bird's eggs or fledglings, and sometimes actively hunt medium-sized mammals such as monkeys or the young of baboons, bushbucks and bushpigs. They follow no set route in their day-to-day search for food and within their home range are nomadic, sleeping close to where dusk finds them. Each chimpanzee constructs his own quite elaborate nest for the night, with the exception of youngsters up to six years, or sometimes older, who share one with their mothers.

Ill. 127

In the wild, the chimpanzee lives in a male-dominated society within which there is a fairly well-defined social hierarchy amongst the adult males. Status is maintained or raised either by means of an actual attack or, more commonly, by means of threat and intimidation. In this respect, the male chimpanzee has developed a magnificent charging display, during which he runs along the ground, sometimes upright, hurling rocks, dragging huge branches, leaping up and swaying saplings. This all combines to make him seem larger and more dangerous than he actually may be.

Chimpanzees communicate by means of a complex repertoire of calls, postures and gestures. Relationships between adult males are, for the most part, relaxed and friendly. Such relationships are maintained, in part, through long sessions of social grooming when groups of males may remain in friendly and prolonged physical contact for well over an hour.

Finally, it is worth stressing that there is tremendous individual variation amongst chimpanzees, both with regard to appearance and behavioural characteristics. I shall illustrate some of the points I wish to make in this chapter by discussing individual chimpanzees, using the names by which we identify each known chimpanzee at Gombe.

Ills. 128. 129. By means of spectacular charging displays, such as this one performed by J. B. (below left), a male chimpanzee may maintain or better his social status without recourse to actual fighting. On the whole, social relationships even between adult males are relaxed. This is clearly demonstrated by the lengthy sessions of social grooming, an activity which permits long periods of friendly physical contact between different members of the community (below right). Here Faben (extreme right) grooms his mother, Flo, who grooms her daughter, Fifi, who in turn is grooming the alpha male, Mike. Hugo and old Mr. McGreggor also groom Mike. A session like this may last for over an hour.

Ill. 130. If males take a tolerant view of the infants in the troop, mothers are possessive about their babies. Baby chimpanzees, like their human cousins, are entirely dependent on their mothers for much longer than other mammals. The necessity to 'educate' an infant and the relatively long time-span of 'growing up' creates bonds which go beyond the instinctive.

BEHAVIOURAL SIMILARITIES TO MAN

Long Period of Childhood. The chimpanzee, like man, has a long period of immaturity. The female does not have her first infant until she is about thirteen years old and the male does not enter the adult hierarchy until he is at least fourteen years old. Thus approximately thirty to thirty-five per cent of a life-span of probably forty to forty-five years is occupied with 'growing up'.

Ills. 130, 132   Most surprising is the length of time during which the child is wholly or
Ill. 131   partially dependent upon his mother. An infant does not take his first
Ills. 133, 135   tottering steps until he is about six months old, and steady locomotion does not commence until the third year. Riding on the mother's back continues to be the main mode of travel until the fourth or even fifth year. Youngsters are seldom weaned completely from the mother's milk until the fifth or sixth year although, from the age of about two years, solid foods become an increasingly important part of the diet. The child usually continues to sleep with his mother until his sixth or even seventh year, or until the birth of a new sibling. One youngster (Flint) continued to share his mother's nest until her death, when he was eight and a half years old.

During the sixth or seventh year juveniles sometimes become accidentally separated from their mothers. When this happens, the youngsters (and sometimes the mothers also) become very distressed and show searching behaviour until they become reunited once more. Subsequently the young male begins to leave his mother, for short periods, of his own accord. Such excursions seldom last for more than a few hours until he is nine or even ten years old. There is a good deal of individual variation in the early attitude of young males to independence. Young females are likely to remain almost continuously with their mothers until they become sexually attractive to adult males for the first time, when they are about eleven years old. Then, during periods of receptivity, they usually travel with males, returning to their families in between.

In many cases, however, mothers die before their offspring reach ten years of age. If this occurs when the child is over six or seven years old he will probably survive, but younger orphans may not, as I shall describe shortly.

Without doubt this long period of childhood is adaptive for the chimpanzee, as it is for man, with relation to learning.[20] In simpler forms of life, much behaviour is almost entirely genetically coded, although at all levels individual experience undoubtedly plays some role in the development of behaviour. As the mammalian brain becomes increasingly complex, however, social learning begins to play a vastly more crucial role. The young chimpanzee, travelling with an experienced and protective female, has ample opportunity to explore and learn about his environment and to watch and learn from the behaviour of his companions, particularly his mother. We know, from experiments in the laboratory, that monkeys and apes are able to learn quite intricate patterns simply by watching another individual performing them.[3] In the wild, it is common to see a young chimpanzee intently watching as another performs some complex activity—such as making a nest, preparing some kinds of foods, using tools or performing a charging display. After watching in this way the child may then try to imitate the behaviour. The importance of the opportunity for such observational learning has been demonstrated by raising chimpanzees in social isolation for the first two years of their life. These individuals subsequently showed a number of the gestures and postures typical of chimpanzee behaviour, but they were invariably used in the wrong social context or appeared as fragmentary and incomplete sequences.[25]

Adolescence. The period of adolescence, often considered to be culturally determined and unique to man, is also an important time in the life cycle of the chimpanzee, both biologically and physiologically.[19] It commences at about nine years of age and lasts for about four years in the female and slightly longer in the male. It represents a further extension of the period of immaturity, perhaps crucial for the perfection of social patterns such as the performance of the spectacular charging display in the male and care of infants in the female.

For the male, adolescence can be stressful. He becomes larger and stronger, his charging display becomes more vigorous, and he directs a good deal of aggression towards adolescent and subsequently adult females who gradually become submissive towards him. At the same time, he must become in-

Ills. 133, 134

Ill. 131. The infant starts to walk at about six months of age. Here Wilkie takes a few tottering steps (opposite). He will not become steady on his feet until he is over eighteen months, and will continue to ride about on his mother's back until he is about four years old. Even then he will jump on to her from time to time, perhaps until he is seven years of age.

creasingly cautious in his dealings with the adult males, who are less tolerant of him now than they were when he was a mere juvenile. Yet he seems fascinated by these mature animals of his own sex. He watches them frequently and his first journeys away from his mother are usually with a group in which there are several adult males. As a young adolescent he occupies a peripheral position in such groups. He may sit some 20 yards from a socially grooming cluster of males, watching them closely. He may display as the group arrives at some food source, but not alongside the other males— somewhere off at the side, in the bushes. He may copulate with a female in such a group, but only after the adult males have done so. Then, when they are quietly resting, he may try to sneak off with the female and, perhaps, mate with her behind a tree, out of sight of his superiors. After a while, particularly if there is aggression in the group, the adolescent male tends to wander away, and travel either by himself for a while or return to his mother.

Eventually, he begins to join in grooming sessions, at first only in the capacity of groomer but finally being groomed himself by an older male. He begins to perform charging displays along with the other males and to show less inhibition in his courtship of females. When he is about thirteen or fourteen years old he starts to challenge the lower ranking adult males and, when he has repeatedly proved himself superior or at least equal to one or two of them, can himself be ranked as socially mature.

For the female, adolescence is usually more peaceful. She is less concerned about establishing her position in a hierarchy and is content to remain with her mother. If the latter has an infant sibling she can learn all she needs to know about maternal behaviour without going further afield, and many young adolescent females do show much fascination for infants, carrying them, grooming them and playing with them.

Ill. 132

Some females are fearful when they first attract the sexual attention of the adult males, since most courtship displays have many components of aggression, but they soon learn what it is all about. And other females take this in their stride.

Perhaps the most fascinating aspect of female adolescence is the fact that a young female, during periods of sexual receptivity, may leave her natal

Ills. 133, 134. Young chimpanzees have a great deal to learn and, for this reason, their long childhood is adaptive. Here, as Winkle grooms Fifi, eight-month-old Wilkie watches Freud playing with a palm-flower 'toy' (above left). Subsequently Wilkie will imitate and practise such patterns himself. Above right: from left to right, Figan, Fifi and Flo all glance towards a sudden noise. Flint, only five months old, has not yet learnt to orientate correctly to potential danger.

community and enter the home range of a neighbouring one, mixing with its members and mating with its males. Usually she returns to her home community when her receptive period is over, but sometimes a female may actually move permanently out of the area where she was raised.

In other species of non-human primates so far studied in the wild it is the young males who transfer from one group to another or leave their own group and try to establish a new one of their own with one or more females 'stolen' from another, thus ensuring the exchange of genes throughout a population.[2,8,31]

In many human societies, it is the woman who leaves her home to move into new surroundings at the time of her marriage. Often, prior to that time, she has led a very sheltered life with her family: suddenly she may have to travel many miles, enter unfamiliar surroundings, even adopt a new culture. Of course, marriages in human societies are the result of traditional customs or love, but if we are interested in the evolutionary heritage which has led to the establishment of human cultures, it is fascinating to speculate that a similar urge to wander, to seek new males, may have existed in our earliest female ancestors. Such an inherent trait would certainly be helpful to young girls when trying to cope with the stressful experience of being 'given' to the son of a family in another group.

The Family. When we compare the family structure in the chimpanzee and man we find one striking difference: in chimpanzee society the father plays no role in the family unit. A male, having sired an offspring, takes no further responsibilities so far as its upbringing is concerned. He may adopt a generally protective role towards *all* young infants in his community, but there is no husband-wife or father-child social relationship. However, between a mother and her offspring, and between some siblings, remarkably close and enduring relationships may develop, as they may do also in the human family. The intensity of these relationships is undoubtedly fostered by the lengthy period of immaturity when the growing child is not only in close and constant association with his mother but also, frequently, with one or more of his brothers or sisters.

Ill. 134

By contrast with most other non-human primate species, chimpanzee mothers are remarkably patient and long-suffering with their youngsters. An infant is permitted to clamber all over his mother's body and she plays with him and grooms him frequently. Weaning is usually a long drawn-out and gentle process, with few vigorous rejections on the part of the mother. She shares food with her infant quite often and may continue to do so even when he has become a juvenile. And, in most mother–infant pairs, there are remarkably few occasions when the mother actually physically punishes her child for misbehaviour: she is far more likely to groom or play with him and thus distract his attention.

Ill. 135

This close relationship persists throughout the juvenile period and, as mentioned, adolescents continue to travel frequently with their mothers, particularly during the early years. Indeed, at this time, when the relationships of the young male with other individuals are constantly changing (as he starts dominating the females and becomes increasingly wary of the older males), his mother probably provides a valuable stabilizing influence. He makes few if any attempts to dominate her and, whilst she may show greater respect for his new-found size and strength, she does not normally show fear of her son.

A chimpanzee mother shows protective behaviour towards her offspring long past the infant stage. On one occasion, for example, a young adult male (Faben) was intimidated by a higher-ranking male and ran off screaming. His old mother (Flo) went charging to his aid, her thinning hair on end, barking aggressively in her hoarse voice. Faben, drawing strength from this support, turned, and together mother and son chased the other male away. When Flo's other son (Figan) was about eighteen years old he sprained his wrist during a status conflict with another male.[11] This happened just a few months prior to his mother's death, when she was very slow and decrepit. Nevertheless, when she heard her son's screams she raced over a quarter of a mile to the scene.[29] She could do little to help but it seemed that her presence in itself calmed Figan, who stopped screaming and limped away through the forest with her. He remained with his family for two weeks, staying away from other chimpanzees until his injury was healed.

In a similar manner, adolescents (including females) may hasten to the defence of their mothers. Sometimes an entire family presents such a united front that it is able to intimidate an individual who could easily subdue each member separately. Thus when Madam Bee, who has a completely paralysed arm, was attacked by the large alpha male, her two daughters (aged about twelve and seven years) ran up and hurled themselves at the aggressor. After a few moments the powerful, aggressive male stopped fighting and moved away.[29]

Brothers commonly assist one another in this manner. Indeed, the young male Figan, who became alpha male in 1973, might never have attained his top-ranking position without the frequent support of his brother Faben. During his struggle for supremacy, Figan seldom challenged another male unless Faben was nearby. And Faben almost always backed up his brother.

Brothers and sisters have not, so far, been observed to spend much time together except when both are travelling with their mother. Brothers, however, associate frequently after becoming independent of their mother and it seems probable that sisters also may spend a fair amount of time together.

Ill. 135. Weaning, for the infant chimpanzee, is a long, drawn-out process which may continue for about two years. The mother is remarkably gentle but, as her milk gradually dries up and she rejects her child more frequently, the youngster may become extremely depressed, playing infrequently, maintaining close proximity with his mother, and reverting to earlier infantile forms of behaviour. Here Fifi, about six years old, rides pregnant Flo.

The old female, Flo, very often travelled with her adult daughter, Fifi, after the birth of Fifi's first infant, so that quite a close relationship developed between old Flo and her grandson. Unfortunately Flo died when he was only one year old.

It seems that there are inhibitions regarding mother-son and brother-sister mating in chimpanzee society. We have observed the behaviour of five physically mature males (ranging from about ten to twenty years in age) when their mothers were in oestrus. Of these, only one made sexual advances to his mother and, whilst she finally permitted him to copulate, she then pulled away, screaming.[37] Flo, during several periods of oestrus, was mated by almost all the males of the community with the exception of her two sons, although they were with her for much of the time. Mating does occur between siblings, but it is very rare. Adolescent females usually try to avoid their brothers if the latter court them during oestrus periods and, after a while, it seems that the males stop attempting to copulate with their sisters. We have not, however, observed enough families to try to relate these seeming inhibitions to any form of incest taboo in human culture.

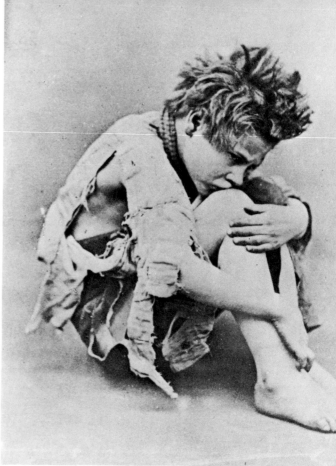

The dependency of family members upon each other may be vividly revealed when a mother dies. Youngsters orphaned when they are younger than five or six years have little chance of survival, despite the fact that if they have an elder brother or sister they will almost certainly be 'adopted'. Three youngsters, all between three and five years of age, showed behaviour closely resembling depression in humans when they lost their mothers. They huddled, lost their appetites, withdrew from most social interactions, particularly play, and became listless and lethargic. One of these was an 'only child' and survived her mother by only three months, during which time she travelled mostly on her own. A second was adopted by a juvenile sister and was often also protected by his adolescent brother. During the year following his mother's death he became increasingly emaciated, showed abnormal behaviour, and finally died one and a half years after his loss. The third orphan was 'adopted' by her adolescent sister, who not only travelled and slept with her, but also allowed the infant to ride about on her back. This child, whilst she too showed many signs of depression, gradually recovered during the year following her mother's death. It is tempting to speculate that the added social security derived by the child from her association with an older and more experienced female, as well as the close physical contact with her, was responsible for the improvement in her condition.

One juvenile male lost his mother when he was about nine. He was, by then, a fairly independent youngster and showed no obvious signs of distress. Two other youngsters of similar age travelled very frequently with

Ill. 138. Juvenile Miff 'adopted' her infant brother, Merlin, after the death of their mother. But, although she travelled with him, and allowed him to share her nest at night, it seems that she was not old enough to provide the security which he needed.

Ills. 136, 137. Opposite: after losing his mother, Merlin became increasingly abnormal and showed many symptoms similar to some of those observed in depressed humans. Merlin frequently adopted this huddled posture, strikingly like that of the orphaned East End boy photographed in the nineteenth century.

their elder brothers after their mothers died and they, too, showed few if any signs of depression. An eight-and-a-half-year-old (Flint), however, became severely disturbed when his old mother Flo died. At that time Flint was still associating constantly with Flo, almost always sleeping with her at night, and even, occasionally, riding on her back.

Flint, who was with Flo when she died, became extremely depressed almost immediately. He only made a few journeys away from the area where his mother died and, for the most part, sat in a huddled posture staring blankly ahead. It is interesting that the only periods of more than a couple of hours that Flint spent away from the general area of his mother's death were the few occasions when he travelled with his elder brother, Figan. On these journeys he was observed sitting close to Figan, sometimes with his hand in contact with the older male. His elder sister, Fifi, spent some time with Flint when she met him five days after Flo's death. But by that time Flint had fallen victim to an intestinal infection and made no attempt to follow her more than a short distance. Three weeks after Flo's death, Flint also died. The illness, striking at a time when he was psychologically depressed and weakened through loss of appetite, took its toll.[35]

Flint, Flo's last surviving child, was unusually dependent on his mother—probably she had lacked the necessary energy to reject his demands on her attention. It is interesting to reflect that the old female's last months would have been extremely lonely had she not been accompanied by her son. During that time she was so frail, so slowed down in her movements, that even her daughter travelled with her far less frequently than before. Whilst one might, anthropomorphically, deride Flint's constant pestering of Flo as he demanded social grooming or pushed at her, whimpering, when he wanted her to move on, it was obvious that the old female became very dependent upon him. Sometimes it was she who turned to follow him if they took different forks of a trail.

Ills. 139, 140. When Flo died at an age of close to forty-five or fifty years, Flint, eight-and-a-half years old, was very distressed. The first day he repeatedly approached her body, as it lay in the stream, and seemed bewildered (below left). Subsequently he became increasingly depressed, showed loss of appetite, social withdrawal and general reduction of all activity. He spent long hours curled up on the ground, staring into space (below right).

Non-verbal Communication. There are a number of striking similarities in some of the gestures and postures of human and chimpanzee non-verbal communication patterns, and in the behaviour of the two species in frightening or anxiety-provoking situations. Both the chimpanzee and man may derive comfort from physical contact with another during such stressful occasions.

If an infant chimpanzee is hurt or frightened he runs to the protection of his mother, jumps into her arms and, often, suckles briefly. This contact with her usually serves to calm him. During the long period of childhood his mother is readily available, and even an adolescent may reach to touch his mother, or hold her hand, if he is suddenly alarmed and she is nearby.

An older chimpanzee, in similar contexts, often reaches to touch a companion, seeming to derive a measure of comfort from this physical contact. Old David Greybeard was even calmed when, after a sudden fright, he picked up and embraced the four-year-old infant Fifi. A similar need for contact is shown by chimpanzees who are suddenly excited. When a kill has been made after a successful hunt, for instance, any chimpanzees nearby may indulge in a veritable orgy of contact-seeking behaviour—embracing, kissing, patting one another and uttering loud screams.

Perhaps the most dramatic illustration of the chimpanzee's need for physical contact is after he has been threatened or attacked by a superior, particularly when a young adolescent male has been victimized by a high-ranking adult male. Once Figan, aged about ten years, was badly pounded by the alpha male (Goliath at that time). Screaming and tense, Figan began cautiously approaching his aggressor, who sat with his hair still bristling. Every so often the desire to flee seemed almost to overcome the adolescent's desire for contact and he turned, as though to retreat. But each time he went on again until eventually he was crouched, flat on the ground, in front of Goliath. And there he stayed, still screaming, until Goliath, in response to his submission, began to pat him gently on the back—on and on until the screaming gradually subsided and Figan sat up and moved away quite calmly. Such incidents are common, and almost always the aggressor responds to the submissive gestures of the subordinate with a touch, a pat or even an embrace. Occasionally, if a young male is not reassured in this way in response to his submission, he may actually fly into a tantrum, hitting the ground and screaming so intensely that he almost chokes.

When two chimpanzees meet after being separated for a while they may ignore each other, one or both may show aggression, or they may greet one another in a friendly way. It depends, to a large extent, on the relationship between them. When greetings do occur, chimpanzees may kiss, touch or pat one another, hold hands or embrace. A male may chuck a female or an infant under the chin. Gestures of this sort indicate the relevant social status of the individuals concerned and undoubtedly serve to re-affirm the subordinate chimpanzee's continuing acknowledgement of the other's superior status. A friendly greeting is often followed by a session of social grooming, the most relaxed and prolonged form of physical contact between adults, and an activity which often helps to reduce tension after stimulating or aggressive incidents.

Humans show a similar need for the reassurance of physical contact. Frightened people may cling to one another. My son, when told of a special treat, often flings his arms so tightly round my neck that I am all but strangled; when peace was declared after the Second World War, people all over war-torn Europe, strangers some of them, danced and wept and sang and embraced one another in celebration. Even the traditionally restrained Englishman may clap his partner on the back when they have brought off a successful business deal—or, at the very least, shake him by the hand.

Ill. 141. Many of the gestures of the non-verbal communication systems of man and chimpanzee are very similar, not only in form but in context, too. Here, for instance, adult male Charlie embraces the orphan, Merlin, in greeting.

Ill. 147

Ill. 146
Ill. 145

Ill. 148

Ill. 142. Opposite: the author searches for chimpanzees in the rugged, forested valley slopes of the Gombe National Park. Grassy ridges between the steep valleys provided good vantage points in the early days of the study when the chimpanzees were so shy of humans.

Ill. 143. Opposite: Flo uses a grass stem to 'fish' for termites. She withdraws her tool from the nest and picks off the termites with her lips.

Family quarrels are often closed with a quick embrace or kiss; a disagreement between friends is frequently terminated by shaking hands; a dejected person may be cheered by the comfort of an embrace or a pat on his shoulders. And human patterns of greeting, in many cultures all over the world, often include the gestures and postures used by chimpanzees. Humans, of course, have lost most of their hair so that they are unable to derive comfort and pleasure from social grooming in the same way as does the chimpanzee. But, perhaps to compensate, man has evolved other forms of prolonged and friendly physical contact, almost entirely lacking in chimpanzee society, such as holding hands or linking arms, stroking the hair, and the elaborate caresses and love-play that precede and often accompany human sexual intercourse.

Many of the patterns and contexts of chimpanzee aggression also bear marked resemblance to some human aggression. One chimpanzee who threatens another may direct vigorous movements of an upraised arm towards him, he may run towards the other in an upright posture waving his arms, he may throw rocks or other objects, often with good aim, or he may brandish a stick. When he actually attacks he may bite, hit, punch or kick. Female chimpanzees, in particular, may scratch or pull out one another's hair. For the most part, however, chimpanzees settle their disputes by threat or by bluff without resorting to actual fighting—the charging display of the adult male represents the supreme example of bluff, since it can make an individual look larger and fiercer and more dangerous than he may actually be. When fighting does occur it is usually brief and, even when it appears ferocious, seldom results in obvious wound or injury.

Ill. 144

Chimpanzees may become aggressive when they compete with one another for social status, or for a favoured food in short supply; when a member of their immediate family is hurt or endangered; in self-defence; when they have been badly startled and are getting over their fright; when they are in pain and apparently 'irritable' (one male, for example, was very aggressive when he had a broken toe). They may be aggressive, too, if they

Ill. 144. Rocks and branches are frequently hurled at random during charging displays. In addition, stones may sometimes be thrown with deliberate aim. Here Flint is ready to throw a rock at Gilka during a juvenile squabble. The chimpanzees sometimes show good aim, but often the missiles fall short of their targets.

Ills. 145–148. Chimpanzees communicate by means of calls, postures and gestures and, in many instances, physical contact plays an important role. In greeting, Mike chucks the female, Passion, under the chin (top) and Melissa kisses Hugo (centre). At no time does the chimpanzee more vividly demonstrate his need for the comfort of touch than after he has been attacked. Here, after being attacked, adolescent Figan has approached his aggressor and, despite his fear of the big male, crouches submissively before him (bottom). In response, Goliath reaches to touch and pat Figan who is soon calmed. Many of the communicative gestures of the chimpanzee show almost uncanny similarity to some of those shown by man (right).

don't 'get their own way': a child who is not allowed to suckle may actually hit or bite his mother; a male, trying to persuade a female to follow him, may attack her if she attempts to escape. A subordinate chimpanzee, who is threatened or attacked or frustrated by a superior, may also be roused to a display of aggression which he usually directs towards some unfortunate individual, lower-ranking than himself, who happens to be nearby. The sight or sound of chimpanzees from a neighbouring community may trigger off displays of aggression, particularly amongst the higher-ranking males, and occasionally 'strangers' are quite savagely attacked.

Through the fourteen years of research at Gombe we have documented some examples of unusual aggressive behaviour in chimpanzees—behaviour which might, perhaps, foreshadow some of the 'inhuman' actions of man.

In their struggle to attain higher status, some males will take advantage of the temporary disability of any higher ranking individual. Figan displayed at and intimidated older males in this way from his adolescence onwards. When his elder brother Faben lost the use of one arm during an epidemic of poliomyelitis Figan, aged about twelve years, challenged him with violent charging displays again and again, until Faben was completely subordinated by his young brother. Figan has remained the higher-ranking brother although, as already mentioned, the two are now close associates and Faben, in fact, assisted Figan in his rise to alpha male.

When old Mr. McGregor was completely paralysed in both legs during the same polio epidemic, the other chimpanzees were initially fearful of his bizarre method of pulling himself along the ground (backwards or head over heels). Soon, however, they became hostile, and a number of the adult males repeatedly displayed at and even attacked the helpless cripple—once shaking him out of a nest he had constructed after pulling himself, with much effort, into a tree.

Eventually, about a week after becoming paralysed, Mr. McGregor was no longer subjected to the violence of his companions—but he was actively avoided by them. Once, he painfully dragged himself some 50 yards to join a group of males who were grooming each other. With a final effort he pulled himself into their tree but, as he approached, the other males quickly swung away and resumed grooming elsewhere. This avoidance was repeated on other occasions until the old male gave up trying to join his former companions. Similarly, when Fifi had a terrible deep wound on her head, the adult male to whom she presented her head, seemingly for the comfort of grooming, repeatedly hastened away from her. Another sick male was also avoided by his companions when he approached, and when the group moved on he was left behind, despite his obvious attempts to keep up.

One adult female (Madam Bee) was repeatedly subjected to seemingly unprovoked attacks by adult males. Moreover, if one male initiated such a fight others present were likely to join in, including females. Such 'contagion' of aggressive behaviour is not at all uncommon in chimpanzee society. Twice, older females, for periods of several months, 'persecuted' two young adolescent females. The aggressors threatened or attacked and chased off the adolescents repeatedly; the latter sometimes started to scream and hastened away when they saw the others arriving. The offspring of the older females sometimes joined forces with their mothers to harass such victims.

Perhaps an additional comment should be made concerning the savage groups attacks which may be perpetrated on individuals of neighbouring communities. Four such incidents have now been observed. Twice, adult females were the victims: both escaped, bleeding badly, but one of them had her two-and-a-half-year-old infant seized by the group of males. The infant was partially eaten by its captors, despite the fact that for almost half an hour it continued to struggle and scream.[4] On the other two occasions the victims were males of the neighbouring community. They were 'caught' near the borders of their home range and very viciously attacked, one by a group of adult males, and the other by three males and a female. Both finally escaped after sustaining serious injuries. One old female was found dead—the nature of the wounds on her back suggested that she, too, may have been the victim of such a group attack.

I should stress that such aggressive incidents are very few and far between. For the most part relationships are friendly and relaxed in a chimpanzee community although, as would be expected, some individuals are far more belligerent, irritable and aggressive than others.

Tool-using. In his use of objects as tools, the chimpanzee comes closer to primitive tool-using performances in man than does any other living creature. It is true that a number of different animal species use objects of their environments as tools, but for the most part the animal is fairly inflexible in both his choice of object and the use to which he puts it. The Galapagos woodpecker finch, for instance, uses twigs or cactus spines to pry insects from holes in the bark—even when he is brought up in captivity with no opportunity for observing this behaviour, he begins to manipulate twigs and spines, pushing them into crevices in bark.[7] By contrast, the chimpanzee uses a whole variety of different objects for a large number of purposes, ranging from food-getting to hostility.

At Gombe, the chimpanzee uses stems and twigs when feeding on termites: he pushes the material into the underground nest, withdraws it and then picks off the clinging insects. In a similar manner he uses larger sticks to raid the nests of two species of ant. He may also use a stick as a lever to enlarge the opening of an underground bees' nest, and as an investigation probe to examine an object which he is afraid to touch or which is too difficult for him to reach. Flint, when his new sibling was born, was repeatedly prevented from touching the baby by their mother: finally he picked a long twig, gently touched his sister with it and then sniffed the end intently. Fifi cautiously touched the head of a dead python with the end of a long palm frond which she then smelt intently.

Ill. 143

Ills. 149, 150

Leaves are used as napkins to wipe dirt or blood from the body—one mother used leaves to wipe faeces from her infant when it dirtied itself. The chimpanzee also uses crumpled leaves as a kind of sponge to sop up water from a hollow in a tree when he cannot reach it with his lips. Old Leakey used such a sponge to wipe clean the inside of a brain case.

Finally, the chimpanzee may use sticks or rocks in aggressive displays. Sticks may be brandished or directed towards a victim (a chimpanzee or an animal of another species) with a clubbing movement. Stones may be thrown, often forcefully and with good aim, either overarm or, less often, underhand. One adult male, Mike, by incorporating empty paraffin cans into his charging

Ill. 144

Ill. 149. Chimpanzees use more objects as tools, for a greater variety of purposes, than any other creature except man himself. Here Nova uses a stick, from which she has peeled the bark, to extract biting ants from their nest (opposite). When the insects swarm up the stick she will sweep it through her left hand and quickly stuff the viciously biting ants into her mouth, chewing fast and furiously.

Ill. 150. Often objects must be modified to make them more suitable for use as a tool. Here Olly breaks a stick, in order to shorten it before using it to try to open a man-made banana box.

displays, so intimidated the other chimpanzees that he became alpha male in about four months—without, so far as we know, any actual fights.

Chimpanzees also modify objects sometimes, in order to make them more suitable for the purpose in hand. Thus, they may strip the leaves off a twig so that it may be inserted into a termite passage, or strip the blades from a piece of grass that is too wide. When they drink with leaves they first chew them briefly, thereby crumpling them and making the sponge more absorbent. Whilst these behaviours are relatively simple, they can nevertheless be considered to be the crude beginnings of tool-*making*. In captivity the tool-making ability of chimpanzees may be quite sophisticated: some individuals can fit three separate tubes together to make one pole long enough to rake in food otherwise out of reach, or pile four or five boxes on top of each other to reach a suspended banana.[15,39]

It seems most likely that tradition plays a major role in the acquisition of different tool-using techniques in different populations, the behaviour being passed from one generation to another by observational learning. Infants spend much time watching their mothers using tools, particularly during the termiting season. One two-year-old watched his mother when she picked leaves to wipe diarrhoea from her bottom: he then picked leaves and wiped his own clean bottom.

We know, too, that at least some tool-using techniques differ from one wild population to another. Chimpanzees in West Africa have been seen using rocks to hammer open hard-shelled food;[1,32] chimpanzees in Uganda may use leafy twigs as fly whisks.[32] Neither of these behaviours form part of the known tool-using repertoire of the Gombe chimpanzees, so that we may speak of them as cultural variations from one population to another.[20]

Hunting and Food-sharing. In his hunting and food-sharing behaviour, the chimpanzee also shows similarities to man. Fifteen years ago, man was the only primate known to hunt in organized groups and it was not generally suspected that many of the non-human primates regularly show meat-eating behaviour. It is now known, however, that chimpanzees in various parts of their range (probably all areas) may hunt, kill and eat medium-sized mammals fairly often.[18,19,34,35] The community at Gombe may kill up to thirty animals in a year.

Most of the actual hunting is carried out by male chimpanzees, although occasionally a female may join in or even make a kill by herself. Sometimes a single chimpanzee will come upon, seize and kill a victim. At other times a group of males, acting in a co-operative manner, may surround, stalk and attempt to capture a potential victim, such as a young baboon a little distance away from his troop. Several times we have observed one chimpanzee creep very quietly up a tree towards the prey whilst the others stood at the base of that tree and also at the bases of other trees which could have acted as escape routes for the baboon. Significantly, even when the chase overhead became tense and exciting, none of the males left their positions until the victim actually made a leap for safety, at which time they converged to try and grab him.

Meat is a prized food for the Gombe chimpanzees. There is much excitement when a kill has been made and the noise attracts other individuals in the neighbourhood. They gather around the male or males who have

Ill. 151. Chimpanzees frequently hunt medium-sized mammals to feed on the flesh. They are the only primates, other than man, in which food-sharing between adults commonly occurs. Here Mike, with the body of a monkey, is surrounded by members of the Flo family begging for shares of the prey.

obtained parts of the carcass and beg for portions. It is in this context of meat-eating that the sharing of food is seen most frequently—other foods are usually found in abundant supplies and there is no need to beg and share. The carcass of a small baboon or baby bushpig is not very large, yet begging chimpanzees are often successful. They may be allowed to feed from the carcass at the same time as the possessor; they may be allowed to detach small fragments; or they may actually be given pieces. Usually a bunch of leaves is chewed with each mouthful of meat and, very often, it is this wad of leaves and meat which is ultimately spat into the outstretched hand of a supplicator. Sometimes, however, a choice tit-bit is detached by the possessor of the meat and placed into the hand of a begging friend. Once Goliath tore in half the body of an infant baboon he had caught and gave half to an older male, who had been throwing tantrums like an infant in his frustration.

One small carcass may be shared by as many as fifteen individuals, though the shares are by no means equal. Old males, who normally rank low in the social hierarchy, are often able to remove large portions of a kill—or even all of it—from younger and more vigorous males. Females in oestrus or individuals such as old Flo, who are persistent and appear to set a high value on meat, also tend to get largish portions or a good number of small scraps. Young adolescent males and juveniles often have to make do with collecting tiny fragments dropped to the ground or licking drops or smears of blood from the branches.

Early man was a hunter, and remains of animals found at his living sites indicate that he gradually became more and more reliant on meat as a food. He developed increasingly sophisticated tools to help in the killing of prey and the cutting up of the carcass. An understanding of the carnivorous behaviour of the chimpanzee gives us a few more clues concerning the evolution of hunting, as a way of life, in early man.

Firstly, since it is during hunting that we see some of the best examples of co-operation in chimpanzees, and in the eating of meat that we see the best examples of food-sharing, it seems certain that more frequent hunting would lead to increasingly sophisticated co-operation and more equal sharing of meat by our earliest ancestors.

Secondly, we can reconsider a frequently discussed question: did early man learn his liking for meat as a result of scavenging from the kills of the predators of his time? All the evidence which has been gathered from the study of the living primates suggests that our ancestors were more likely to have initially hunted for themselves. Chimpanzees, it is true, may sometimes snatch a baby bushbuck from a baboon who has just killed it, but apart from that we have never observed them feeding on any meat not killed by a member of their own community, even when they had the opportunity to consume freshly killed carcasses. Baboons, on the savannah, frequently hunt for the young of gazelles or antelopes,[12] but I have found only one anecdotal report of baboons scavenging. And that was not from the kill of a predator, but the body of a small antelope killed by a car on the road.

It is not surprising that the primates of today do not attempt to scavenge from the kills of lions, leopards or cheetahs. Even hyenas and jackals, scavengers *par excellence*, have to be extremely wary; anyone who has watched, as I have, the deadly charge of a lion or the swift spring of a leopard will realize the danger inherent to the scavenger. In those far-distant days when early man began to venture into the dangerous savannah, he was probably armed with nothing more than a stick or a handful of stones. Even today, when most animals have learnt to fear and avoid man, I would not care to join a small band of people, armed in this way, and try to drive a lion or, worse, a hunting group of hyenas, from their rightful kill.[22]

I am certainly not trying to suggest that early man never scavenged. Man is, and undoubtedly always has been, an opportunist. If he saw the chance of scavenging safely he would surely have done so—after he had become more reliant on meat as a source of food through his own hunting efforts.

From the above descriptions of chimpanzee behaviour we cannot, of course, assume anything about the social structure or behaviour of early man. But, it seems probable that behaviour common to modern chimpanzees and modern man appeared in our ancestors, and that there were enduring affectionate relationships between mothers and their offspring and between at least some brothers and sisters. The old saying 'blood is thicker than water' may have its roots way back in our earliest human forebears.

We can suppose that early man had a long childhood, during which he explored his environment in play and learnt the traditional behaviours of his group by watching others and imitating and practising things which they did. It seems probable, too, that early man went through a period of adolescence during which, perhaps, the male gradually took his place in the hunting excursions of the adult males of his group whilst the female probably spent her time with other females and may have 'helped' her mother by playing with and carrying around her younger brother or sister.

When early men were frightened or anxious they almost certainly would have reached to each other for the reassurance of contact. After fighting they undoubtedly made it up (if they made it up at all) with a reassuring pat or clasp of hands or an embrace. When they met after a separation they may have kissed, embraced or held hands. Until that point in time when they lost their hair, they probably groomed one another in the typical primate way: as their hair receded, perhaps, more and more often they made grooming-like movements such as stroking or patting, or sat holding hands for companionship.

It seems likely, too, that early man used the grasses, leaves, sticks and rocks of his environment as tools long before he ever chipped the first flake or bone to make a durable artifact. Probably he continued to use the original primitive tools along with the new and more sophisticated stone ones for a long period of time. To this day, the Bushmen of the Kalahari and at least one tribe of South American Indians use leaf sponges in the same way as do the chimpanzees. And anyone will pick up a long stick lying nearby to investigate, for instance, a seemingly dead snake.

Finally, we can suppose that, with increased reliance on the savannah and the hunting of mammals for food, early man became more sophisticated in co-operation within his group and shared food more and more frequently.

PRECURSORS OF CHARACTERISTICS UNIQUE TO MAN

An understanding of chimpanzee behaviour not only reveals some remarkable similarities between these apes and man, but also helps to highlight man's uniqueness, to pinpoint some of the characteristics that set the human species apart from other primates. Man has evolved a distinctive upright posture, the female is constantly sexually receptive and, at some point in his history, man lost most of his hair. More significantly, the explosive development of the human brain and human language has led to sophistications in understanding and reasoning that dwarf the sensitivities and the intellect of even the most gifted chimpanzees.

Can we see, in chimpanzees today, patterns which might be precursors of these uniquely human characteristics? In some cases we can, and through an understanding of chimpanzee behaviour and some of the major trends in human evolution, we can speculate as to how such shadowy precursors might have evolved towards the distinctive human form.

The Upright Posture. First, let us consider the bipedal gait. Chimpanzees sometimes adopt an upright posture, mainly when they are trying to see over long grass, when they are performing aggressive displays, when they are carrying food and when they are travelling over wet ground. Most individuals are awkward when they move upright, tending to bend forward at the waist and showing a waddling gait, and they seldom travel thus for more than a few steps at a time. Some chimpanzees, however, are able to maintain an unusually upright posture, with few signs of waddling, over quite long distances. Two individuals (one being Flo's son, Faben) learnt to travel upright for many yards at a time after each losing the use of an arm during the polio epidemic, thus preventing the useless arm from trailing along the ground.

What environmental pressures might, conceivably, select for more frequent and more efficient bipedal locomotion in chimpanzees? If a group began to move out into the open savannah as part of their daily routine, into an area where predators were common and trees few, it would be a distinct advantage for a chimpanzee to be able to maintain an upright posture easily. He would be able to look out over long grass and consistently spot dangers, or potential prey, ahead. He would be able to locate his companions more easily if, for some reason, he became separated from the main group. After a successful hunt, the chimpanzees would probably prefer to take the carcass back to the forest to eat in the safety of the trees. For they eat meat very slowly and, moreover, like to stuff in a handful of leaves with each mouthful, and such material is often unavailable on the plains. An individual who could carry his share whilst walking upright could more easily see a lion or a hyena approaching, or some other creature which might try to snatch away his meat. Even if the group went out searching for berries or other vegetable foods quite close to the forest, they might like to gather an armful to take back to the shade and safety of the trees to eat. Such a journey would only be worthwhile (before the invention of a carrying vessel) if *both* hands were free for carrying the food.

Finally, encounters with other predators on the savannah, especially around a kill, would place a premium on the effectiveness of the intimidation display. The chimpanzee often performs this in an upright posture as he waves branches or hurls rocks at an adversary. Many of the most dramatic upright displays of this kind at Gombe have been during aggressive encounters with baboons. If chimpanzees had to intimidate another species of ape-like creatures which also displayed in an upright posture, there would be a distinct advantage for the individual whose performance excelled.

Ills. 152, 153. For the most part the chimpanzee walks on all four limbs. He can, however, stand and walk or run in a bipedal position. Often an upright posture serves to intimidate others during an aggressive incident, as when Circe swaggers threateningly at subordinate Melissa during banana feeding (opposite). Some infants repeatedly try to walk in an upright posture, as Flint does here (above).

Ill. 144

Most major evolutionary developments are undoubtedly shaped by a complex of *different* factors which, working separately, ultimately lead, through natural selection, to a common end in the genetic or cultural change in the behaviour concerned. There is one other factor which might be considered with regard to the evolution of bipedal locomotion: the inability of an infant primate to cling to its mother.

This may have been due to the development of a larger brain and the consequent necessity for the baby to be born before the head became too large to pass through the pelvis. In this case, the newborn infant may have been too immature to cling. Or it might be due to the loss of hair, so that there would be nothing to which the infant could cling. A mother would then have to support her infant with one hand as she carried it around. If she was able to walk upright she would still have one hand free for collecting food and other tasks. Thus females who were best able to walk upright might well be the most efficient mothers and, over time, their genes might be increasingly selected for. (Incidentally, this inability of the infant to cling to

Ill. 154. Sometimes a chimpanzee stands or walks bipedally when carrying food, when trying to see over the long grass, or when looking for a lost companion.

its mother would virtually necessitate the adoption of an almost entirely terrestrial mode of life, for the adult females at least.)

It has frequently been argued that bipedal locomotion was necessitated, in early man, for the efficient use of tools. The chimpanzees at Gombe, however, perform virtually all of their tool-using behaviours in a sitting posture, which frees the hands. Only when man had developed sophisticated artifacts which he needed to carry with him from place to place would the upright posture become really advantageous for tool-use although, as already discussed, it is essential for the efficient use of weapons.

Constant Sexual Receptivity. Of all the living species of primate, it is only the female of *Homo sapiens* who is continually sexually receptive to the male throughout her reproductive life and beyond the menopause. In other species, mating typically occurs during the period of heat or oestrus. (Occasionally, when monkeys or apes are kept in captivity in small cages, a male may mate with a female when she is not in oestrus, but this must be extremely rare in the natural habitat.) The sexual cycle of the higher primates is characterized, as in the human, by a period of menstrual bleeding and, in some species, by a pronounced swelling of the sex skin which coincides with oestrus. The length and frequency of these cycles varies from species to species: in the chimpanzee it lasts about thirty-five days, during which the female develops a shell-pink swelling of the sex skin and is receptive to males for about eight to ten days.

The chimpanzee is characterized by a great variety of, and lack of rigidity in, sexual patterns. Males have a number of courtship displays, the one used depending on their age, their individual idiosyncrasies, or the female they are courting. The postures adopted by both partners during copulation are flexible. And the behaviour of the female during her oestrus varies too, depending partly on her individuality, partly on her age, and partly on the male or males she is with. Sometimes a female may be mated by all the available males throughout her period of genital swelling. As many as seven males may copulate with the same female within seconds of each other. Usually a female goes off with one male during the peak of oestrus—the time when conception is most likely to occur.[23] This may be a different male in each cycle, but sometimes she goes off with the same male during several successive periods of oestrus. This is not to say that the male remains 'faithful' to this female in between her cycles, nor that she will refuse the advances of other suitors. However, it suggests that there is an incipient tendency towards more stable pair-bonding.

Some females are a great deal more sexually attractive to a larger number of males than are others. Such popular females may be either young or old. One young female at Gombe (Fifi) showed what might be considered the chimpanzee equivalent of nymphomania: during her early periods of swelling she constantly sought out males and solicited copulation, even to the extent of manipulating the penis of one old male until he showed an erection, got up and mated her. Fifi even continued to solicit males for a couple of days after her first adult swelling had completely subsided.

After the female attains menarche, around ten years of age, there is typically an infertile period lasting between one and two years. During this time the young female cycles regularly, becoming well integrated into male society. After conception, young females usually continue to show fairly regular swellings, at which time they are receptive, and often attractive, to males. Such swellings may continue until the sixth month of the eight-month pregnancy. These young females may commence to cycle again when their first or second infant is about two years old, even though they do not, apparently, become fertile for another two years.

Thus chimpanzee females of under some twenty years of age lead fairly active sexual lives. As they get older, however, their periods of sexual activity are likely to become widely separated. Older females are unlikely to become sexually receptive during pregnancy, nor are they likely to start cycling until their infants are about four and a half to six years of age. Then, within a few cycles, they are likely to conceive again and forego sexual activity for a further five or six years. These older females spend a great deal of time in association with their own offspring only, whereas younger females are much more often with groups of chimpanzees which include adult males. One apparently sterile female has been cycling regularly since menarche: she is now about twenty years old and still associates with adult males very frequently.

Let us now return to the unique phenomenon of constant receptivity in the human female. Most theories concerning the evolution of this sexual characteristic have centred around that period in prehistory when early man had established division of labour—when the men hunted and the women stayed at home to gather berries and other vegetable foods. Constant sexual receptivity, it has been argued, was a necessary attraction to ensure that the hunters returned to the women to distribute the meat.[26] It is equally possible, however, that factors contributing to increased receptivity were operating at an earlier stage of man's history, when his way of life was, perhaps, more similar to that of the chimpanzee today.

What factors might operate to produce longer and more frequent periods of receptivity in female chimpanzees? As we have seen, older females—those who are sexually inactive for long periods—tend to spend increasing amounts of time away from social groups. If chimpanzees began to move frequently into the relatively more dangerous environment of the savannah, females travelling on their own might fare worse than those moving with adult males, since it is the latter who show dramatic intimidation displays in response to danger. Moreover, females travelling with males would be more likely to get a share of the meat after a successful hunt. And, of those females who *were* there, those in oestrus would often be most likely to get good portions. We can suppose, then, that if meat began to play an increasingly important role in the diet, those females who showed more frequent and longer periods of receptivity would have an advantage over others and, in time, their genes might be selected for.

Also of relevance to this discussion is the fact that, in some instances, psychological or social pressures may directly influence sexual receptivity in the female primate. In Hamadryas baboon society, each male sets himself up as leader of a small group of females. A young male typically initiates his 'harem' by taking a one-year-old female from her mother. He acts in a most paternal manner towards her, carrying her across difficult routes and protecting her from danger. In this species, the female develops her first sexual swelling and becomes receptive to her male a whole year earlier than do the females of other baboon species.[17] This precocious sexual development is, of course, a characteristic that has become, as a result of natural selection, part of the genetic heritage of the Hamadryas female.

If we turn to the chimpanzee, we find numerous instances in which physiological and probably psychological stress can *inhibit* sexual swelling. But there is one instance where psychological factors seem to have led to the *development* of a swelling in one female—the old mother, Flo. She was grooming with several adult males when a young pregnant female with a large pink swelling suddenly appeared. The males left Flo and hastened to groom the newcomer and to inspect (by touching and sniffing) her swelling. For a few moments old Flo, whose infant was just under a year old, stared at the young female. Her hair was on end, as it is in an aggressive chimpanzee. Then Flo slowly approached the group and carefully inspected the swelling which had attracted the males. The following day Flo had developed a small sexual swelling—just sufficient to attract the attention of some of the adult males so that they went to groom and inspect her. It was the first swelling that Flo had shown since her infant's conception. By the following day it was gone and she did not swell again until her infant was four and a half years old.

I have described this single incident because any information concerning an increase in receptivity in female chimpanzees may be of interest to those trying to understand the unique human condition.

Whilst the ability to sexually attract and satisfy a member of the opposite sex from puberty to old age has, without doubt, had some influence on the evolution of man as we know him today, it is only a part of the bond which may be formed between adult humans. Man is capable of a great depth of feeling for another that may be quite independent of blood relationships: he has a capacity for loving with passion yet tenderness, with sympathy, understanding, patience and joy.

In chimpanzee society there is normally no close relationship between non-related adults, although the affectionate bonds between a mother and her offspring and between siblings may be strong and enduring. As we have seen, such bonds may have been strengthened, during evolution, by the need for a longer and longer childhood, which led to family members becoming increasingly dependent on one another for physical needs or emotional comfort and security.

Once evolution has provided a species with a new characteristic it can be adapted, or modified, to perform different tasks to that for which it was originally selected. Man did not develop an upright posture in order to delight others with the beauty of the ballet: cutting-tools were not evolved for the carving of a Michelangelo statue. So, when evolution had equipped our ancestors with the capacity for forming meaningful, affectionate attachments within the family, it would have become possible for similar bonds to be developed between non-related individuals. If, through time, the close bond with the mother continued to deepen, a young male, as he matured and became increasingly separated from his mother, or after her death, might search for a substitute, a continuation of the emotional security of his relationship with her.

When early man was living in small groups at the edge of the savannah and the males went on hunting forays, whilst the females stayed closer to the living sites with their infants, stronger bonds of affection between opposite sexes would be advantageous. The male would be more likely to save a good portion of the kill for a female, and she more likely to keep a supply of roots and berries for a male, if they were emotionally attached to one another.

Why, I wonder, should it be thought that constant sexual receptivity must have been a prerequisite for the establishment of sexual-bonding man? Other primates form stable sexual bonds. The gibbon associates with one adult female only, probably for life. She is receptive to him during oestrus—a few days each month—unless she is pregnant or lactating, which she is for much of the time, in which case he has no sexual activity at all. The male Hamadryas baboon has several females in his 'harem' but there are many months when none of them are sexually available to him: he does not seek satisfaction with his neighbour's females, nor does he abandon his own.

Suppose we turn the usual hypothesis around: the development of an *emotional* attachment between males and females might have had some effect on the willingness of the female to mate with her male—even a desire, on her part, to attract him. And, as the act of procreation became increasingly tied to the new affection between the sexes, so it would be pleasurable in a new way and sought after more intensively.

Intellectual Capacity. Man's capacity for loving his fellows, for sympathizing with and trying to understand them, is only one of the traits which sets our species apart from the rest of the animal kingdom. Man is aware of himself, as an individual and as a species. He is aware, and becoming ever more aware, of his relationship to the rest of nature and the part he plays in the world around him. He knows that he can control, to some extent, the events of his life, that he can make crucial decisions that will affect his future. But, because of circumstances beyond his control, such as inexplicable misfortunes, or unexpected good luck, man has, from time immemorial, tended to acknowledge the presence of supernatural forces. He praises or blames fate, or the god or gods of his religion. Man knows, too, that he is finite. And because he values life, he clings to the belief of immortality for his spirit.

Man's ability for abstract conceptualization is so sophisticated that it seems almost pointless to search for precursors in his closest living relative—the evolutionary gap is so vast in this respect. Yet man's intellect did not blossom overnight. In our earliest ancestors there must have been faint glimmerings of dawning consciousness; sparks which flamed to full human realization during the amazing evolutionary development of the human brain. This is the fabled 'missing link' of anthropological folklore—not a strange half-ape, half-human skeleton, but a series of vanished brains, each more complex than the one before. Brains that are for ever lost to science save for knowledge of their relative size and a few faint imprints on fossil craniums; brains that held, in their increasingly complex convolutions, the dramatic serial story of developing intellect that has led to modern man.

If, therefore, we seek to learn more about this incredible progression, a closer examination of the intellectual capabilities of the chimpanzee may, in

Ills. 155, 156. A spectacular form of charging display may be performed when a male chimpanzee comes upon a racing stream or one of the magnificent waterfalls that plunge down some of the rocky Gombe stream beds. These unique photographs show Figan displaying at the base of the Kakombe Falls (below) and, as he becomes increasingly stimulated, swinging wildly on hanging lianas growing beside the Falls, high above the ground (opposite). Did similar emotions underlie the beginnings of prehistoric religious beliefs?

fact, be worthwhile since, as I have already mentioned, the structure of his brain is more similar to that of man than is that of any other living creature. Moreover, compared with most monkeys, he has a highly developed cognitive ability. His complex brain has permitted the development of the correspondingly complex behavioural repertoire which I have briefly outlined. In the natural habitat, however, his full intellectual potential is seldom realized, although we get glimpses of it—in Mike's use of empty paraffin cans to raise his status, or in the tactics shown by Figan in his rise to the alpha position. But it is in captivity, where intelligence can be spurred to greater performance by the offer of rewards, that we can learn most about the intellectual powers of the chimpanzee.

Ills. 195-197 In the past few years, some young chimpanzees have been taught to communicate in the sign language of the deaf or by using plastic symbols for words.[10,28] They learnt quickly and were able to string signs or symbols together in meaningful ways. Washoe, for instance, is able to communicate in sign language such sentiments as: 'Please take me into the garden to see flowers'. When told, in sign language, that there was a big, bad dog outside who ate little chimps, her eyes grew large and round; when asked, 'Does Washoe want to go outside?' she emphatically shook her head. Sarah, using plastic symbols, showed that she had an understanding of abstract concepts: she was taught that 'yellow is a colour', 'blue is a colour', and for herself was able to reason that 'green is a colour'. Another young female, Julie, is able to guide a bead (using a magnet over the glass cover) through an incredibly complex maze after spending but a few moments, initially, working backwards from the end point.[30] And, lest it be thought that I am gunning for superior female intellect, a young male, Sultan, showed outstanding ability in solving problems using reason.[15] These were mainly problems involving the use of objects as tools. Sultan worked out a variety of solutions to obtaining food lures out of reach, including the fitting together of three poles to make a tool long enough to reach the reward.

Chimpanzees are well able to discriminate between different shapes and colours and they are readily able to transfer information from one sensory modality to another—an object which they have learnt to recognize by sight alone they can then recognize by touch, in the dark. Finally, home-raised chimpanzees who have been treated like human children (or usually rather better!) develop a truly remarkable repertoire of behaviour in which their powers of observation and imitation are strikingly revealed.[13]

As we progress up the evolutionary ladder and learning comes to play an increasingly important role in the development of behaviour, we find correspondingly greater differences between the performances of individual animals. This is strikingly so in the chimpanzee and, in particular, there are marked variations in the intellectual abilities of different individuals. In other words, there are gifted chimpanzees and exceedingly stupid chimpanzees. It is worth noting that the sexual pattern is probably adaptive in relation to the propagation of genes linked to intelligence. In many monkey societies it is the alpha male of a troop who typically sires most of the infants. In a chimpanzee community all the adult males have the opportunity to sire young, so that a chimpanzee of outstanding intellectual ability, even though he may not be motivated to acquire alpha status, has the opportunity to pass on his genes.

Ill. 157. Opposite: Flint starts to climb at about six months, keeping close to his mother. If he ventures too far away, or seems to be in difficulty, she will rescue him.

Self-awareness. The chimpanzee is aware of himself, although not with the brilliant clarity of man's self-awareness. The young female Washoe was shown her reflection in a mirror. When asked (in sign language), 'What is it?' Washoe, after feeling the mirror, looking behind it and into it, signed back, 'Me, Washoe!' Other experiments with mirrors show that chimpanzees, unlike monkeys, are able to recognize daubs of paint, on their mirror images, as belonging to themselves. They even use mirrors to investigate parts of their bodies that they cannot otherwise see.[9] But no chimpanzee, surely, *knows* that he is aware of himself. He sees 'through a glass, darkly', and has not yet come face to face with the realization of himself, a chimpanzee in a chimpanzee world.

Altruism. As we have seen, a chimpanzee typically responds to the submissive gestures of an anxious subordinate with a gesture which serves to calm and reassure, just as a man may cheer a friend with an embrace or pat on the shoulder. Does this then imply the precursors of altruistic behaviour in the chimpanzee? It depends on our definition of altruism. Very often, in man, the proximity or the sight of an unhappy person disturbs another's sense of well-being. It troubles him so that he feels compelled to try to comfort. The sight or sound of her child in distress is intensely disturbing to a mother: part of her motivation in alleviating his distress is to calm her own anxiety. It may well be that the sight and sound of a tense and screaming individual crouching before him is disturbing also for a chimpanzee. His reassuring touch may serve the dual purpose of relieving his own feeling of unease in addition to calming the subordinate.

Is much of man's altruistic behaviour merely a way of soothing his own disturbed equilibrium? Undoubtedly it is in many cases, yet this in no way detracts from the inspiring nature of human altruism at its best: how incredible that the very *thought* of starving people, thousands of miles away, can so disturb a person that he is willing to make a personal sacrifice of time or money to try to help the unknown victims. That a man may go out of his way to rescue and tend a wounded stray cat because he cannot bear the thought of it suffering. The chimpanzee has not yet evolved to the stage where he would put himself out for any individual other than a family member or some other very close associate. (By the same token, no chimpanzee would plan the killing of countless unknown chimpanzees for the sake of personal glory or economic gain. His intellect is neither capable of soaring to the same heights nor sinking to the same depths as that of man.)

Religion. It makes no sense to talk about 'religion' in relation to the chimpanzee. Yet, if we seek to learn more about prehistoric man's dawning awareness of supernatural powers that seemed to rule his life, there is one chimpanzee behaviour which is of interest. At the onset of very heavy rain, or at a sudden loud thunderclap, or when the wind sweeps the mountains with near gale force, hurling dead palm fronds to the forest floor, male chimpanzees may perform spectacular charging displays. Sometimes these are identical to the displays seen when a male arrives at a good food source, or when two groups meet, or during status conflicts. Often, however, the 'elemental' display is of longer duration and seems to have a more rhythmic quality. Goliath, for example, performs a display in the rain that is quite different from his usual

Ill. 158. In response to a sudden torrential downpour, a howling gale, or a sudden clap of thunder, a male chimpanzee may perform a wild, rhythmic and spectacular charging display. Here Figan shakes a sapling as part of his impressive elemental display (opposite).

Ill. 158

162

charging display, during which he runs fast over the ground, hitting at vegetation and hurling rocks. In the rain, his display is more magnificent as he charges, in slow motion, from one tree to another, stamping and slapping on the ground, pausing to sway branches to and fro, to the accompaniment of the rustling of the vegetation and the pounding of the rain. Once I saw a whole group of adult males displaying together in a torrential downpour, charging down the green mountain slope one after the other, only to walk back up and charge down again, tearing off and brandishing large branches as they did so.

Perhaps it is even more significant that the fast-flowing water of the valley streams may elicit such displays: for the streams are no sudden, unexpected happening but are always there, in the same place. Yet time and again a male, who has been peacefully wandering through the forest, sometimes by himself, will pause to display along the stream bed, hurling rock after rock as he does so. Hidden deep in the mountains are several spectacular waterfalls, some with sheer drops of a hundred feet or so, and here, too, swaying on the vines that hang beside the racing cascades, the chimpanzees may perform *Ills. 155, 156* their displays, on and on, watched perhaps by birds and monkeys, all sounds drowned by the roaring thunder of the ever-moving water.

Is it not possible that the first faint stirrings of awe and wonder, that underlie most religions, might have had their origin in emotions similar to those impelling the chimpanzee to perform his wild elemental display? From such primeval, uncomprehending surges of emotion, man's awe and wonder would have increased as his awakening intellect became more aware of the significance, to himself, of the natural phenomena around him. Wildly hypothetical as this may be, it is certainly true that many of the gods and godesses of primitive religions are symbolized by natural phenomena, and countless people have worshipped water as well as the sun, the moon, the stars and all the other wonders of nature.

Death. A chimpanzee does not know that he will die and he does not bury his dead. Yet he does seem to have some ability to learn about death—to learn that death is the end of living. This became apparent when I watched the reaction of two different females to the loss of their infants. Both mothers carried the bodies of their dead babies around with them for a few days, as is the custom of most non-human primates. One of the mothers was a young female and the infant had been her firstborn. Even when the three-month-old baby had been dead for some time she continued to treat the body with the same solicitous care that she had shown during its life. She cradled the corpse when she sat, she pressed it to the correct position against her breast when she travelled. The second mother (Olly) was much older, and experienced in child-raising. Almost certainly she had lost a child before, and possibly more than one (since she only had two surviving offspring prior to this baby, with a gap of about ten years between them). This newest baby was one of the first victims of the polio epidemic and lost the use of all four limbs. For some two hours I followed as Olly travelled up the valley with *Ill. 159* her sick infant. Her progress was slow, as she stopped to embrace him whenever he screamed, cradling him with care, arranging his limp arms and legs so that she would not crush them, pillowing his tiny head in her hand and gazing down at him.

The baby died whilst Olly sat, grooming her daughter, during a heavy rainstorm. And when Olly moved off she showed a complete change in her treatment of the infant. Now she slung the body carelessly over her shoulder and, if she sat, the corpse was permitted to drop to the ground with a thump. She picked it up by one leg, slung it over her shoulder and moved on. This sudden change of attitude was completely different from the gradual diminishing of maternal care that occurs in other mothers—a diminishing that seems to be related to the increase in decomposition and odour of the corpse, so that it resembles, less and less, an infant chimpanzee.

Olly's young daughter, however, had not learnt about death. She played with the corpse, groomed it, and even pulled a dead hand towards her ticklish neck and laughed. For the past week she had been trying to touch her little brother and had been prevented from doing so by their mother. Now she had her chance.

Language. If the chimpanzee shows so many traits that might be similar to behaviours that led, in man, to love, self-awareness, altruism, religion and awareness of death, why has he not, in the past millions of years, evolved to a higher level of consciousness? Or, to put it another way, what was the spur that urged our own ancestors onwards so that, today, there exists such a tremendous evolutionary gap between human intellect and that of our closest relatives?

Of course many factors were involved: biological, environmental and social, each entwined with and affecting others in a complicated tangle of evolutionary progressions and dead ends that can never be wholly clarified. But there is one attainment which, above all others, stands out as a unique milestone in human evolution, that differentiates man from the chimpanzee in kind as well as degree—human language. Without his language, without the ability to formulate and communicate ideas with words, to discuss events of past and future, to pass on, by word of mouth, the wisdom of past and present to the next generation, the forebears of *Homo sapiens* might never have progressed beyond the evolutionary status reached by the chimpanzee today.

Chimpanzees do have a complex communicative system based not only on the postures and gestures which have been briefly discussed, but also on a large repertoire of calls, each one specifying the context in which the sound is made and the identity of the individual making it. For instance, if a juvenile is attacked or hurt he screams; his mother will identify his calls and hurry to his protection, though she will ignore the screams of another youngster. But neither by his calls nor by his gestures can a chimpanzee convey anything other than events which are happening in the present or which have happened in the immediate past: 'I am eating good food' or 'I have just been hurt'.

Why, in view of his considerable intellectual ability, has the chimpanzee not developed a spoken language? Recent research into the brain suggests that the calls of chimpanzees are not the direct precursors of human verbal language, although they are very similar to the laughs, shouts and screams of our own species. Even with the most intensive and loving training, infant chimpanzees, brought up in the home from the first day of life, have not learnt to utter human words, except for one who finally made vague approximations of 'Mama', 'Papa' and 'Cup'.[13] This is because, as recent research has shown, a neurological link between sound production and the cognitive, intellectual centres of the chimpanzee brain has not been developed.

Thus it appears that our own spoken language is a completely new development, involving a different area of the brain altogether from that which is concerned with the production of non-verbal sounds. What circumstances might have arisen to provide the stimulus for developing a more precise mode of communication in our early ancestors? The question has been asked over and over again: the hypothesis most frequently quoted argues that an increased reliance on hunting would provide an incentive.[38] However, whilst the hunting way of life would certainly demand more sophisticated co-operation and, eventually, tool-use, some of the most efficient hunters in the world achieve success without verbal language. Predatory mammals such as wolves, hunting dogs and lions are excellent hunters. The Australian aborigines and the Kalahari Bushmen actually suppress the use of speech whilst hunting and use, instead, an elaborate series of signs—the kind of 'language' which, in fact, we can imagine chimpanzees using much more readily than words.

Without doubt, a whole complex of different factors was operating at the time when human language dawned, each contributing its own stimulation;

166

each, as it were, giving the brain yet another boost towards the same goal. Co-operation—particularly in deciding when and where and what to hunt—was probably one of those factors. As affectionate attachment between individuals and intellectual ability both increased, the desire to share new experiences with one another may also have been crucial. There is another factor which might have played a significant role: the evolution of an infant primate which was no longer able to cling to its mother.

Let us ask how such a development might affect communication in the chimpanzee. The infant is in close contact with its mother during most of its first year and signals between the two are mainly tactile and visual. The growing baby is, of course, exposed to chimpanzee calls but very few of them are actually directed towards him. His mother, during social interactions with others, may call or respond to calls. The infant may then join in with calls himself, but he is not *required* to do so as an aid to his survival. Moreover, as mentioned, mother chimpanzees tend to spend long periods of time away from other chimpanzees so that, for hours or days, the world of the infant may be virtually without chimpanzee vocalizations.

If, however, an infant was not able to cling to his mother, there might be occasions when he was laid down to enable her to use both hands for food collection, the making of nests and so on. And this would presumably lead to an evolutionary new need for vocal exchange between the pair. A frightened chimpanzee infant screams, quietening only when his mother (or another) gathers him into an embrace. A mother whose infant responded to her *voice* by quietening would have a distinct advantage over a mother who always had to leave her task to gather up her noisy child. (This would have been even more important for our own ancestors when food preparation had become more complex and there were other tasks requiring the use of both hands.)

If the infant was deprived of the constant reassurance of physical contact with his mother, *verbal* contact might become a substitute. If he made a little sound and his mother responded in kind, he would feel less lonely. Thus we have created a situation in which, for the first time in the evolution of the primates, a high premium would be placed on exchange of vocal signals between a mother and her infant. This, of itself, would not lead to language. But is it not possible that it might have far-reaching evolutionary implications for the development of the vital neurological link between cognition and sound production?

Nowhere do we see more subtle and constantly changing patterns of communication than between a mother and her growing infant. Many, if not all, of the communicative signals of adult chimpanzees may have been derived, in the evolutionary sense, from those seen in the mother–infant relationship. Just as this bond is more intense and more durable in the chimpanzee than it is in the monkey, so it would have become even more significant in our own ancestors. It seems logical to me that factors leading to human language may have been embedded in this relationship, the heart of all social relationships. Perhaps the equivalent of 'Mamma', 'Maman', 'Mama', 'Mum', 'Mummy'—the word is remarkably similar in many languages—was one of the first spoken words of human speech. I have suggested just one factor—there are many others which could have become more significant with the evolution of the human intellect.

CONCLUSION

All theories concerning the appearance and way of life of our earliest ancestors are speculative to a large extent. Certainly we have true evidence of his bony structure, but no comparative anatomist can tell us the colour of his eyes or skin, whether or not he had hair and what kind of hair, whether he tended to be thin or fat. We find primitive stone tools in association with the remains of animals and we can surmise, as Louis Leakey has done so vividly, the uses to which these tools were put. But we cannot be certain. We cannot even be sure that all of the animals whose bones we find there were, in fact, killed or even eaten by man.

The ideas in this chapter, however, are even more speculative—concerning as they do early man's vanished social behaviour and the environmental and social pressures which may have helped to shape the evolution of an ape-like ancestor towards modern man. We have no time-machine, we cannot project ourselves back to the dawn of our species to watch the behaviour and follow the development of our forebears; if we seek to try to understand these things a little we must do the best we can with the flimsy evidence available.

I have based much of my argument on the hypothesis that man and chimpanzee diverged, in the remote past, from a common ancestor. If we accept this theory, it seems reasonable to suppose that those behaviours shown by modern man and modern chimpanzee did occur, in similar form, in our Stone Age ancestors. The concept of early men poking for insects with twigs and wiping themselves with leaves seems entirely sensible. And, for me, the thought of these ancestors greeting one another with kisses or embraces, reassuring one another, begging for a share of food, is appealing. The idea of close affectionate ties within the Stone Age family, of brothers helping one another, of teenage sons hastening to the protection of their old mothers and teenage daughters minding the babies, brings the fossilized relics of their physical selves dramatically to life.

Even if we do not accept the concept of a common ancestor, we cannot fail to be impressed by the close likenesses in the biology and behaviour of man and chimpanzee today. Familiarity with the behaviour of a creature which is, in many ways, so similar to ourselves certainly highlights vividly the uniqueness of our own species. The fact that we find behaviour patterns in the chimpanzee which might be similar to those which, in early man, developed into human characteristics—such as love, altruism, religion—should be, at the very least, helpful in stimulating further thinking as regards the evolution of human behaviour.

Why does man, today, act in the way he does? In love and hate, violence and altruism, progress and destruction? The answers are buried in a tangled complex of factors in which biological, environmental, social and moral strands are interwoven. And these strands stretch back into the mists of man's origins, linking him, inescapably, with his past. Linking him, through his genes, to a time when behaviour patterns (such as aggression, hunting animals, need for surrounding space), which may seem maladaptive in modern society, probably enabled Stone Age man to survive.

It is important to try to understand modern man—indeed, the future of the human species may depend upon our doing so. Insight into the behaviour of our remote ancestors may be helpful: at any rate, we cannot afford to reject any line of enquiry that shows promise. In addition, it is vital that men of reason collaborate in their efforts to unravel the factors that make us act as we do. Biologists must join forces with philosophers, sociologists with ethologists, anthropologists with geologists. The Knights of the Round Table eventually discovered the Holy Grail through the efforts of one of their number. But man is more complex, more elusive, than any Holy Grail of his imagination. The Knights must join forces in a collaborative attempt if the quest is to be successful.

Ill. 161. The Bushmen of Botswana, although not 'primitive' men, live a hunting-and-gathering existence similar to that of early man. It was during this phase of man's evolution that he developed the behavioural traits and characteristics which we have inherited. It is therefore interesting to study the way of life of the Bushmen to see if it is possible to extrapolate from it some of the traits of early man. Here, two Bushmen, crouched on the ground, aim their arrows during a hunting expedition. If they are successful, they will distribute the prey amongst their families and close friends—a behavioural trait which is prefigured in the meat-sharing of the chimpanzees (see *Ill. 151*).

7 The Bushmen

Irenäus Eibl-Eibesfeldt

THE behaviour of the Bushmen offers scientists a unique but rapidly vanishing opportunity to study one of the few hunting-and-food-gathering societies which has survived into the twentieth century. Man is known to have lived at the economic level of the hunter and food gatherer for the greater part of his history, and it is believed that any phylogenetic shaping of behaviour could only have taken place during this lengthy period of his development. The period since the invention of agriculture is generally considered to have been too short to have exerted any significant change in his genetic constitution. The question as to what extent man's 'nature' is predetermined by the history of his genetically inherited adaptations has led a number of anthropologists and ethologists to initiate studies of Bushman behaviour during the past decade. Existing hunting-and-gathering societies provide good models for the behaviour of early man, but it has to be remembered that all such societies are not 'primitive' but modern, that is, adapted to recently existing conditions. The societies themselves are not 'primitive' but their organisms and cultures may exhibit more or less primitive *features*—patterns which, once established, did not change during the course of evolution.

It must be emphasized that by referring to hunters and gatherers one does not refer to a uniform strategy of life. Actually, a significant adaptive radiation exists. Eskimo hunters, after all, exploit their environment in a different way from Bushmen. They share certain aspects of social life, in particular by living in small communities bonded on an individual base, in contrast with the anonymous societies of Western civilization. Economically, however, they lack domestic animals, with the exception of dogs, and they do not cultivate plants. Their material culture is often said to represent a palaeolithic level. These are certainly 'primitive' features.

Of the recent anthropological publications, those of Marshall,[18] Tobias,[32] Heinz,[9] Lee and DeVore,[17] and Silberbauer,[29] as well as Traill's,[33] are particularly noteworthy. The Harvard Study Group, under Lee, is actively engaged in an extensive research project on the !Kung* in Botswana. A research group for Bushman ethology is, at the present time, composed of H. J. Heinz, an anthropologist who has worked with the Bushmen since 1963, D. Heunemann, H. Sbrzesny and myself. The last three participants have worked with Bushmen since 1970, visiting them regularly every year. We concentrated on two bands of one nexus* of the !Ko Bushmen, which used to live close to Takatswane in Botswana, but have recently moved to Bere, also in Botswana. Furthermore, we have also started to document the G/wi in the Central Kalahari and the !Kung of South-west Africa.

* See p. 184 for notes on the transliteration and pronunciation of the Bushman language.

* Bushmen live in bands composed of several families. Several bands form an alliance system called the nexus system (see p. 181).

INBORN BEHAVIOUR PATTERNS

The documentation programme encompasses social interactions, rituals and cultural skills, so as to cover as many of the facets of Bushman life as possible and thus learn in which ways specific cultural patterns contribute to their survival in the ecological niche that Bushmen occupy. What are the laws by which the cultural patterns are shaped, to what extent are patterns of behaviour shaped by function, and, finally, which patterns prove, in cross-cultural comparison, to be universal?

This latter question is of particular interest, since it refers directly to the question of the nature of man. It is here that we may expect to discover patterns inborn in man as phylogenetic adaptations. This does not mean that all universal patterns are innate. A pattern could be brought about by similar environmental influences acting, for example, upon the individual during

Ills. 162–165. Kissing, which is still a universal token of affection, finds its origins in kiss-feeding. Kiss-feeding, when food is passed from one mouth to the other, is a behavioural trait which man shares with chimpanzees. Opposite, an !Ko Bushgirl is shown kissing her younger sister. After contact with the lips, she pushes a morsel of food into the baby's mouth. The friendly and affectionate sharing implied in kiss-feeding is continued in the practice of kissing on the mouth or cheek when no food is passed (right).

early childhood and shaping its behaviour. There are, however, cases where such formative environmental influences can be excluded by probability,[6] and where the cross-cultural similarity is explained by a common genetic heritage. Numerous examples have been described in recent years. Some of the patterns seem to have a fairly old heritage since, by comparison with higher primates, homologous patterns can be found. Our Bushman research has provided a number of examples.

One of the patterns shared by man and chimpanzee, for example, is the kiss. Rothman and Teuber first drew attention to the fact that chimpanzees *Ill. 146* on friendly encounter contact each other in a mouth-to-mouth, kiss-like fashion. Often, but not always, food is passed. This *Zärtlichkeitsfüttern* is *Ills. 162, 163* derived from mouth-to-mouth feeding as employed by mothers when feeding their young with premasticated food. Jane van Lawick-Goodall observed *Ill. 141* that chimpanzees in nature greet each other by kissing and embracing. Kiss-

173

feeding, furthermore, was observed in orang-utans and gorillas,[2, 20] and it has been repeatedly suggested that the human kiss is homologous with the behaviour of the anthropoid apes.[2, 20, 22, 37]

Our Bushman studies provided evidence supporting this hypothesis. I filmed the act of kissing performed by the Bushmen many times, and frame-by-frame analysis allows us to conclude that kissing was derived from mouth-to-mouth feeding. During kiss-feeding, food is pushed into the partner's mouth with the help of the tongue, but the tongue movement is also observed during kissing itself, when no food is passed on. *Ills. 164, 165*

Other striking examples concern expressive behaviour. The 'play face' or relaxed open-mouth display, as it is also called, occurs in chimpanzees and man as well. We filmed it often in Bushmen children. The same is true of pouting. It is a pattern of submission and withdrawal. The offender often tries to re-establish the interrupted bond. Frequently, the sulking child covers his head with one or both arms. This is interpreted as being derived from the original function of protecting the head against blows, a pattern which is also found in chimpanzees. *Ill. 167*
Ills. 166, 168
Ills. 169, 170

To provide one last example: in a Bushman society, sexual presentation occurs in a primate-like fashion. There are a number of forms of dancing during which the apron covering the rear is jerked to the side in order to expose the buttocks, and this, by the way, is also seen in some European dances. At the ritual of girls' initiation, there is a special dance (Eland dance), which serves as a fertility ritual,[7, 24] during which the buttocks are uncovered. It must be mentioned in this context that the buttocks serve as a strong sexual signal for the Bushmen and therefore are normally kept covered, except for such ritual occasions.

The buttocks of the Bushmen and Hottentots are very pronounced, because of the strong steatopygia which can be interpreted as a signalling device (releaser). It appears, from Stone Age paintings and carvings, that *Ills. 171, 172*

Ills. 166–168. Another behavioural trait which the Bushmen share with chimpanzees is the 'play face', a relaxed, open-mouth display (left). This can be seen clearly in the photograph on the far left, when an eleven-month-old !Ko Bushboy tries to hit his father with a stick, and on the right, when he playfully hits a friend.

Ills. 169, 170. Pouting is a reaction to a stare threat, a form of aggression in Bushman societies. Above right: an !Ko boy uses the stare threat to try and intimidate the girl, who pouts. In the picture to the right, the offender's hand is seen touching the girl's neck, in an attempt to re-establish contact. This is similar to the reassurance patting of chimpanzees (see *Ill. 147*).

Ills. 171–173. Steatopygia is a feature which is common among Bushmen and Hottentots, and this physical trait must have originated from very early times, as shown below in the limestone fertility goddess figurine from Willendorf, Austria, which dates from the Upper Palaeolithic period. In Africa, the Queen of Punt, brought back to Egypt by Queen Hatshepsut's expedition, also shows the symptoms (left). The bulging buttocks have become a signalling device for the Bushmen, and the Cul de Paris, depicted in this detail of Seurat's *Sunday Afternoon on the Ile de la Grande-Jatte* (opposite), seems to show that emphasis on this part of the anatomy still has appeal.

this feature was originally more widespread, even on the European continent, and indeed a preference seems to have survived the reduction of this signal, since European women in fashion still emphasize this region (Cul de Paris). This evidently attracts the attention of men.[6,37]

During mocking, genital displays, including sexual presentation in a primate-like fashion, are performed by girls. They approach the person they want to tease, often lifting their genital apron. Upon close approach they turn around and, exposing their rear to full view, bend deeply. Since the lordosis is particularly well developed in the Bushmen, the vaginal orifice is completely exposed to view and the enlarged *labio minora*—another particular feature of the Khoisan—becomes functional as a signalling device.

For copulation, Bushmen still prefer insertion from the rear, whereby the partners lie on their sides, the buttocks of the woman pressing against the abdomen of the man. Certainly, we are dealing here with one of the older primate behaviour patterns, which persist in some human groups more than in others.

TERRITORIALITY AND AGGRESSION

In a number of recent publications, it has been put forward that hunters and gatherers in general, and the Bushmen in particular, are especially non-aggressive and do not show territorial behaviour but, in contrast, live in open groups with a shifting membership.[10,23,38] This being so, it has been argued that man was non-aggressive and non-territorial by nature, and only changed his peaceful ways with the so-called 'Neolithic Revolution', when he began to cultivate land and fight over its possession. It has been further

176

Ills. 175, 176. When mocking, Bush-girls perform genital displays, including sexual presentation in a primate-like fashion (above right). This can be paralleled by the 'naughty' pose of the can-can dancers (above left).

Ill. 174. Opposite: a Bushwoman and her child in front of their crude shelter made of branches; this protects them from the sun and the seasonal rains. Although Bushmen live in egalitarian societies, personal possession of bows and arrows, clothing, trinkets and household goods, such as the enamel bowl in the picture, is allowed.

stated that chimpanzees, too, live in open groups and are particularly non-aggressive. This all sounds very nice and plausible but unfortunately it is simply not true.

Careful anthropological studies do not show that the majority of hunters and gatherers are particularly non-aggressive, nor do recent studies of chimpanzees indicate particularly non-aggressive habits. On the contrary, the rich repertoire of appeasing patterns clearly indicates a strong aggressive potential which has to be checked, and indeed Jane van Lawick-Goodall describes aggressive displays, attacks against members of the same species who are strangers and even against members of their own group who, after an epidemic of poliomyelitis, showed deviant behaviour[13] (see pp. 147, 148). Japanese authors recently reported that chimpanzees live in residential groups, each separated from the other by a strip of no-man's land. Gorillas have been shown to live in small, closed groups spacing each other by aggressive displays.

And what is the situation concerning the allegedly peaceful hunters and gatherers? The groups particularly mentioned in this context are the Eskimos, the Hadzas, the Bushmen, the Pygmies and the Kwakiutl. These certainly do not constitute the majority of hunting-and-gathering societies and thus, even if one could confirm that they are especially peaceful, one could not say that this is a trait typical of hunters and gatherers in general. The Ona Indians, the Andaman Islanders and some Australian tribes, for example, are known to be rather aggressive hunters and gatherers. However, not even the above-mentioned allegedly non-territorial and 'remarkably' non-aggressive peoples prove to be so on closer examination. The Kwakiutl Indians, described by Schjelderup[27] as people lacking the 'instinct' to fight, are well known for potlach feasts, during which the chiefs compete fiercely to outdo the guests by destroying valuables. Their songs are aggressive and they call the performance a fight,[1,3] and they not only destroy valued objects in the competition for prestige, but also kill slaves. Surely such activities can only be described as acts of aggression?

As for the Eskimos, it is again difficult to see why they are considered to be particularly non-aggressive in face of the rich variety of aggressive acts that have been described. The tribes of Siberia, Alaska, Baffinland and North-western Greenland settle their disputes by wrestling and occasionally a wrestler gets killed. Eskimos in Central Greenland slap one another's face, and in Western and Eastern Greenland song duels are a favourite means of settling disputes. Rasmussen[21] described numerous aggressive acts. It is a fact of interest for the history of science that we know, in the case of the Eskimos, how the myth of the aggression-free society was created. According to König,[11] it was Nansen who, wanting to create a favourable image of the much persecuted Eskimos, depicted them as a harmless people, horrified by the rude, aggressive ways of the Europeans.*

Regarding the often quoted Hadzas, who, according to Woodburn,[38] defend no territories and live in open groups, it must be said that he studied them at a time when they were displaced from their original territory, measuring about 5,000 square km., into an area of approximately 2,000 square km. In fact, Kohl-Larsen,[12] who studied these people in 1934–6 and later in 1937–9, recorded the ritualized ways in which groups fight.[5] And from the allegedly peaceful Pygmies, fights between clans and also territorial ownership by bands have been described.[25, 26]

All these facts are, however, omitted from the discussion, and many people continue to believe in the non-aggression of hunters and gatherers, in spite of the vast amount of evidence to the contrary. It would almost seem that the less some of the promoters of the neo-Rousseauian myth are hampered by knowledge, the more outspoken they are. Some people make sweeping statements about hunters and gatherers, and occasionally go so far as to write whole books on their ways of life, without ever having seen one, even from a distance![28]

Confronted with this controversy, I paid particular attention to the questions of territoriality and aggression in the Bushmen. Sahlins[23] and Lee[15] have written of Bushmen as living in open, non-exclusive bands and of not being territorial. Concerning the !Kung, Lee reports:[15]

> The camp is an open aggregate of persons which changes in size and composition from day to day. Therefore I have avoided the term 'band' in describing the !Kung Bushmen living groups. Each waterhole has a hinterland lying within a six-mile radius which is regularly exploited for vegetables and animal food. These areas are not territories in a zoological sense, since they are not defended against outsiders.

There are, however, numerous references in other works which clearly document evidence to the contrary.* Passarge[19] describes the !Kung Bushmen as belligerent and he stresses that not only the bands but every family owns its own collecting ground. Zastrow and Vedder[39] emphasize that the Bushmen are not allowed to hunt or collect food in another band. Lebzelter[14] reports that the !Kung Bushmen show great distrust when they meet members of another band and that two men who approach each other will lay their weapons on the ground. Among others[4,34,36] who also report on territoriality, Marshall[18] stresses that the territories are shaped in a general

* Nansen is the only source on which this opinion is based. In fact, he learnt very little of Eskimos in their natural state and his moral point of view was tendentious in order to arouse sympathy—as he himself stated in his *Eskimo Life*.[11] A thorough examination of the literature provided additional evidence of Eskimo territoriality in particular, and territoriality and warfare in hunters and gatherers in general. These facts will be presented in my forthcoming book, *Krieg und Frieden*, Piper, Munich.

* In fact, in a recent publication, Lee[16] does point out that his !Kung Bushmen are quite aggressive and have a particularly high murder rate. In addition he refers, in contrast with his previous publications, to ownership of waterholes and land, thus implicitly correcting previous statements concerning territoriality. Polly Wiessner (verbal communication) in addition proved the existence of nexus systems (see p. 181) in these !Kung.

178

Ills. 177, 178. A family of Kalahari Bushmen leaves for the hunt (above). Women carry the children, while the men hold their long spring-hare poles and digging sticks. During the hunt the women will gather wild berries and root vegetables. The rock-painting from Battle Cave, Basutoland, (below) shows Bushgirls going to collect vegetables, carrying their digging sticks in much the same way.

way around the patches of *veldkos* which are 'jealously' owned, and Tobias[32] writes that:

> Territoriality applies among the bands of the same tribe and between different tribes. Tribal bounds are sometimes reinforced by social attitudes, such as the traditional enmity between the Auen and Naron. Under special conditions, such as an abundance of food, these bounds and the accompanying enmity are forgotten.

Probably this latter situation characterizes the group of Bushmen studied by Lee. Indeed, he repeatedly emphasized the abundance of Mongongo (mangetti) nuts. His Bushmen—again in contrast with other groups—experience no seasonal shortage of food. Certainly the !Kung Bushmen do not normally live in open and non-territorial groups, and, according to Silberbauer,[29] neither do the G/wi. Referring to the ethological definition of territory as 'a space in which an animal or a group generally dominates others, which become dominant elsewhere', Silberbauer writes that this 'aptly describes the relationships between G/wi bands with regard to their territories; a visiting band, or a single visitor, submits to the dominance of the host band either by waiting for an invitation or by seeking permission to enter and occupy the territory'. He emphasizes, furthermore, that both visitors who pass a territory en route to another destination and those who are headed for the occupying band call at the band encampment and ask permission 'to stay in your country and to drink your water'. This is, by the way, a standard phrase, which is even used when the waterholes are completely dry—as is the case during most of the year. In each band, individuals are addressed as owners (!u:ma, or !u:sa in the case of a woman), who are, according to tradition, the founders of the band or their male or female descendants.

Silberbauer refers to the band as being 'open', relating to the fact that visitors, after asking for permission, may come and stay for a while and that occasional adoption into a band occurs, although he stresses that the change in group composition is 'not very frequent'. However, the correct ethological term for this is a 'closed group' (which is never absolutely closed, as adoptions occur according to rules) to distinguish it from an 'open group', where no group membership exists, since the members change freely—as, for example, is the case in most schools of fish.

Last, but not least, I would like to bring to attention the numerous Bushman cave-paintings found in South Africa, which depict armed fights between Bushmen bands, as well as inter-ethnic fights between Bushmen and Hottentots or Bantus.[31] The latter occur when the Bushmen raid the cattle of the other tribes.

Ill. 179

Territoriality in the !Ko Bushmen. Working with the !Ko, our research team thoroughly explored territorial behaviour. Heinz[9] has made it clear that the !Ko distinguish three levels of social organization. Firstly the family, secondly the band, and thirdly the band nexus. All these units have definite patterns of bonding *and* spacing. Rules regulate how family members build their huts in clusters, in which order the members sit around the camp-fire, etc., and though all band territory is accessible to every member, preferred areas for the families' activities are recognized.

Ill. 174

Ill. 179. Opposite: a rock-painting showing a rustling expedition by a tribe of Bushmen. An armed battle is in progress between the owners of the cattle, the Bantu, and the raiders. The former pursue the Bushmen, who defend their prize without the benefit of the shields carried by the Bantu.

On the band level a definite territorial claim exists. Control over it is exercised by the head-man on behalf of the band.* Hunting, collecting of wood and *veldkos* regularly takes place within the band's territory only. In case of emergency, the territory of an allied band (belonging to the same nexus) may be used, but only after formally seeking permission from the head-man of the neighbouring band.

The highest organizational level is represented by the nexus. This is an alliance of several bands, which are connected by numerous marriage ties.* They share peculiarities in dialect, visit each other to perform certain rituals and refer to themselves as 'our people'. The nexus territories are demarcated from each other by a strip of no-man's land. Each territory is exclusive and no member of another nexus would dare to ask for permission to hunt there. Access to territory is acquired by birth or marriage. If the parents come from two different bands, dual membership results. The groom who resides for a while with the bride's band has access to its territory, and in turn the bride acquires the right to collect in the territory of her husband when the couple move there.

Undoubtedly the !Ko are territorial and this is certainly no exception in Bushmen societies, as our previous discussion has shown. It is to be hoped that this fact will soon be recognized by the advocates of non-territoriality,

 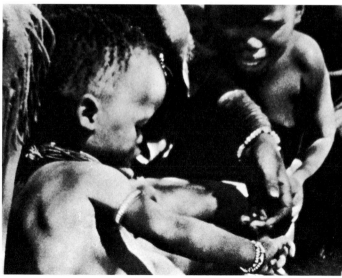

particularly since, for the sake of their ideological concepts, they are willing to deny the Bushmen their right to land ownership, on the grounds that, if they have never been territorial, they have no justified claim for the land they occupy.

Aggression in Children. It is recognized by most students of man that an aggression-free society does not exist. However, it is argued that warfare, or sorcery as the ritualized form of aggression, is not known in Bushmen and that aggression generally plays a minor role in the life of these people, who are primarily friendly. This in turn is explained by the 'fact' that they experience little frustration in childhood, particularly since they are allowed intensive bodily contact with their mothers. Let us examine these points.

Observations in the !Kung and !Ko have revealed that the relations between siblings are regularly very tense, particularly during the first year of the younger sibling's life. The newborn regularly gets most of the mother's attention and the older one, boy or girl alike, is much less cared for by the mother, who pushes the child away. The change in the attention with the birth of the sibling is most dramatic, since the mother–child relationship is a very intimate one.

Relationship of the siblings is characterized by numerous aggressive acts. The older sibling tries to push, kick and scratch the younger one and he teases him by taking away his toys. The younger, in turn, attacks either spontaneously or in response, and from seven to eight months of age is very effective in defending his place at the mother's breast. The relationship between siblings improves with age and finally, after a phase of ambivalence, a lasting friendly bond is established.

Babies from the ninth month onwards are very possessive; they try to steal each other's objects, defend them and also their position at the mother's breast. Their patterns of attack are highly functional: the opponent is pushed over, scratched or slapped. They are also very efficient in handling sticks for attack or retaliation. These patterns develop, although Bushmen do not instruct their children to be aggressive or to retaliate once attacked.

Ills. 180, 181. Sibling rivalry is just one of the forms of aggression between Bushman children. Here, the older of two !Kung brothers tries to scratch his younger brother. His mother prevents him from doing so by taking his arm away and he cries with frustration.

Ills. 180, 181

Ill. 182

182

Aggressive interactions are a common occurrence in the playing groups of children from four to twelve years of age. Playful aggression—characterized by laughing, and the fact that no spacing or submission results from the encounter—can be distinguished from serious aggressive interactions during which the children slap, kick, bite, scratch or box each other and *Ill. 183* threaten or throw objects at each other. These result in spacing, cut-off behaviour and/or submission (including crying). Such aggressive interactions are remarkably numerous. On one occasion, when a group of seven girls and two boys were playing together, I counted 166 acts of aggression and defence during an observation period of 191 minutes.

Although aggressive acts are numerous, aggression is neither encouraged nor rewarded by the approval of group members. The older children, in particular, are active in comforting the distressed, punishing the attacker and encouraging friendly patterns of bonding by sharing food and initiating play dances. The socialization of aggression, indeed, takes place within the play

Ills. 182, 183. Aggression does not only occur between siblings. Above right: an !Ko boy pushes over a little girl who has tried to take a toy away from him. Below right: an eight-year-old !Ko Bushgirl is about to throw a stick at a boy who has teased her.

groups and with very little interference from the adults. The oldest girl of a playing group observed in Takatswane (!Ko) dominated the younger children for two years, initiating and guiding the play activities, instructing the smaller children and keeping quarrelling at a low level. She intervened when a quarrel broke out and used to scold and punish the aggressor. She was not peaceful, however, and often initiated attacks, seemingly to maintain her dominant position.

Many of the quarrels were about the possession of objects. Often I had the impression that smaller children were trying to find out what others would tolerate. Among the patterns of aggressive behaviour, the ritualized stare threat deserves particular notice. The children often used to stare at each other until finally one gave up, by averting the eyes, lowering the head and pouting. The pattern seemed to be universal. Mocking and teasing are also common forms of ritualized aggression. (I described them in detail and discussed their function in my monograph on the !Ko.[5]) Bushmen have a pronounced inclination to tease those who deviate from the norm, and even persons whose behaviour is changed by disease are ridiculed. Children fear strangers. They flee from them or push them away in active repulsion. Stranger hostility was found to be a universal pattern which developed even if the child never experienced any harm from strangers.[6]

Aggression in Adults. Between adults, aggression is often verbally expressed. Miss Liz Wiley, who worked as a teacher in Bere for almost two years, kindly provided me with a number of typical phrases that she noted when a fight broke out between members of two neighbouring bands of the same nexus. The following verbal insults were uttered (I use the traditional form of writing):*

> *a maga'i*—you shit!
> *a sa a tshxa*—go and eat shit!
> *n ||aba kane ka a*—I don't want you!
> *a ba n|'a*—you will die!

Women often use animal references:

> *a ki n|u*—You are a hyena!

Insulting phrases often refer to sexual organs:

> *a ≠n ≠gaba |i*—your penis is bad!
> *a |anate ≠auku be chuny*—your leaves (labia) are as long as those of a baboon!

From Dr. Heinz I got the following phrase of insult:

> *Ke ≠ a'a ≠au ku bi ≠uli*—your clitoris is like a long stump!

The phrase '*n Ki dzai ma a e*' translates literally as: 'I am hungry, I will eat you', meaning: 'I am so angry with you that when I fight you, I'll eat you!'

Early this year the two !Ko bands near Bere were engaged in a fight. A man had approached a married woman from another band of the same nexus, whom he had courted many years ago. The brother of the woman intervened. Others soon took sides in the quarrel, according to band membership, and the men fought. As a result, several people were injured: one man got an open wound on the head, another a broken rib, and the hand of a third

* The Bushman language contains many clicks. Five symbols are used in the anthropological literature to write them: ! is pronounced by sharply pulling the blade of the tongue off the roof of the mouth; ⊙ is pronounced by giving the noise of a kiss; | is pronounced by drawing the tip of the tongue with a sucking noise sharply from the rear of the front teeth; || is pronounced by clucking, as if encouraging horses; ≠ is pronounced by withdrawing the tip of the tongue sharply from a point just behind the front teeth.

man was so seriously crushed by a bite that he needed hospital treatment. Smaller wounds from scratches and bites were numerous. The women only joined in the quarrel verbally, their main concern being the separation of the fighters.

Additionally, I learnt from a Bushman about a case of sorcery which occurred recently. A member of the Takatswane band had been insulted by a Bantu. He retaliated by throwing bones with the intention of causing harm.* To my knowledge, this is the first record of sorcery. It is, however, known that the !Kung used a small bow-and-arrow set as a magical weapon. The bow and the arrowhead were carved from oryx horn, and a bow which I collected measures 19·5 cm., while the arrows are 16 cm. long. (These became favourite tourist souvenirs and they are now mass-produced and carved from wood, but stained black to make them appear to be made from horn.)

Lebzelter[14] reports observations of Dornan that the Kalahari Bushmen used to dance before engaging in war. On this occasion they would shoot little poisoned arrows into the direction of the enemy or against the sun, as it is said they believe that the shadow of the enemy is killed by this action. Originally, they did not like to part from their magic bows since they were afraid they would get into the hands of their enemies. Vedder[34] also refers to these bows as magical weapons. According to him, the Bushmen shoot their arrows in the direction of their enemies while, at the same time, uttering curses.

According to Heinz,[7,8] males tease and maltreat underdogs within their band. Verbal threats of murder are often uttered, such as: 'I will kill you with my medicine!' This again refers to sorcery, but he does not state exactly how the threat would be carried out. He also describes the temper fits which occasionally overcome the Bushmen:[8] 'An angry Bushman finally settles down with a face that shows an unbelievable degree of anger. It takes very little for this anger to cause a wrestling and punching encounter with sticks and knobkerries. If the reasons are serious the fight will deteriorate into one in which knives and spears are used. . . .'

I think these observations make it evident that Bushmen are certainly not particularly non-aggressive people. However, they only use their aggression to socialize according to their pacifist ideal. Aggression is discouraged rather than encouraged in children, and Bushmen generally avoid conflict. Thus families which are endangered leave the band for a while instead of fighting. It is customary for the insulted party—in the evening, when everyone can hear—to complain loudly about the affair. He does not refer to his opponent by name but everyone knows who is meant and, since the social pressure is very strong in such a small group, the offender will always approach the offended during the next few days to placate him and initiate friendly contact again.[5,7]

EGALITARIANISM AND POSSESSION

It is occasionally argued that hunters and gatherers live in egalitarian societies and that individual property plays no significant role. Such simplified statements do not adequately depict the reality.* It is true that in small societies, where people know each other well and intimately, no rigid hierarchy develops. Nevertheless, people strive for esteem and they are

* Bone-throwing is also used by the !Ko and G/wi as an oracle. From the way bones are scattered, the Bushmen read where they should go for a hunt. In a similar way they ask the bones if, and from which direction, visitors might come. The !Kung use round pieces of oryx-skin for the same purpose.

* Referring to the gender role, Schmidbauer[27] writes: 'Far-reaching equality connected with *rigid division* of labour characterizes the contribution of the sexes with regard to the survival of a Bushmen group'. However, it has in fact been observed that women do the bulk of work by collecting food, building huts, fetching firewood and bearing the burden of child-rearing.

admired and appreciated for their particular virtues. A good hunter and a good dancer, for example, derive satisfaction from their performances. Old people, men and women alike, are furthermore respected for their experience and knowledge. Founders of a band and their offspring are considered to be the owners of a territory. Headmanship exists and sometimes, though not always, follows the line of the founder of the band. The band is often referred to by the name of this person, for example 'Midum's band'.

As far as property is concerned, Bushmen definitely own their personal items such as bow and arrow, clothing, household goods, trinkets and the *Ill. 174* like. Ownership is strongly felt and some objects are only shared with a selected group of people, mostly close members of the family.[7] Owned objects play an important role in everyday life. There is a strong group pressure in favour of sharing and giving, within fairly strict rules, by which bonds are established and reinforced. Jealousy and envy play a most important role in an equalizing process, but it is cultural pressure that works along this line and it should be understood as a special cultural adaptation of the Bushmen.

The same holds true for the sharing of game. The hunter is considered to *Ill. 161* be the owner and has the right to distribute his prey, again following strict rules. The different parts of the game are distributed according to closeness of relationship (compare with chimpanzees, pp. 151, 152). In the !Kung, the matter of ownership is more complicated. The owner is the person who owns the arrow that hits, and this introduces an element of chance, since hunters sometimes use arrows of other band members. If the hunter successfully kills with another person's arrow, the latter will be the owner of the prey. This, however, is a secondary way of abolishing the importance of ownership in order to avoid rivalry and conflict.

CONCLUSION

Ethological studies of the Bushmen have revealed a number of basic behaviour patterns which can be interpreted as phylogenetic adaptations in man, some of which are shared with some non-human primates. As hunters and gatherers, Bushmen represent an archaic type of human economic and social organization, and research has shown that many behavioural patterns are, in cross-cultural comparison, universal to early man and, indeed, to present-day man. I firmly believe that there is indisputable evidence of territoriality and aggression in Bushmen, thus indicating that men fought over the possession of land long before the advent of food production. At the same time, however, sharing, both of objects and game, does take place in Bushman society, despite the strong importance attached to ownership of personal items, and this would seem to suggest that the instinct of sharing may not have been completely dormant in early man.

The Mind of Man

8 Evolution and Man's Self-Image

Theodosius Dobzhansky

Ill. 185

Ill. 188

Ill. 12

SCIENCE has radically changed man's ideas of the world he inhabits. Perhaps most of all, science has altered man's self-image. Philosophers of classical antiquity believed in eternal permanence and fixity of the cosmos. Of course, everyone observes changes going on in himself and in his surroundings. Such changes were declared delusions of the senses. Plato affirmed that human beings, and all material entities, are mere shadows of the everlasting and immutable Ideas. Wisdom consists of discerning these Ideas underlying their transitory embodiments.

The germ of the evolutionary world view appeared within the Judaeo-Christian religious philosophy. Man and the world are neither unchanging nor involved in tedious cyclic changes and returns to the same old states. History is pregnant with meaning. The divinely ordained historical process had a start in Creation, passed through stages of the Fall of Man, revelation of religious truth, and redemption, and is progressing towards God's Kingdom. History is full of hope; there may be a better world and better men. How ironic that the opposition to the evolutionary idea came in modern times chiefly from religious circles!

Until a century ago, the idea of evolution was hamstrung by a huge underestimation of the duration of the historical process. The world was believed to have been created as recently as five or six thousand years ago. No major changes could plausibly be assumed to have taken place in nature or in man within so short a time. The world was consequently believed to have always been approximately as we find it today. Moreover, it was taken for granted that the world was created specifically for man, who received it as his domain.

Aristotle, and after him Ptolemy, posited a stationary earth located at the centre of the universe. Around the earth were revolving several concentric crystalline spheres with heavenly bodies embedded in them. Nearest to the earth was the sphere containing the moon, next followed the spheres of the sun, of the then known planets, and finally the sphere of the 'fixed stars'.

The so-called medieval synthesis, a world view resting on biblical and Aristotelian cosmologies, was expected to endure permanently. But the Aristotelian cosmology caved in when Nicolaus Copernicus, born half a millennium ago, replaced the geocentric model of the universe with a heliocentric 'hypothesis'. As Goethe rightly said, 'Of all the discoveries and opinions proclaimed, surely nothing has made such a deep impression on the human mind as the science of Copernicus.' Not only was the earth dethroned from its presumed centrality and pre-eminence, but Copernicus

Ill. 184. A model of the double-helix structure of the DNA (deoxyribonucleic acid). DNA is the carrier of the genetic information transmitted from the parental to the progeny genetrations.

189

correctly inferred that the universe was much vaster than hitherto imagined. The 'fixed' stars are far more distant from the Earth than is the sun. Kepler, Galileo and Newton, during the two centuries following Copernicus, not only corroborated the heliocentric model but made it the foundation of a new cosmology. Newton brought together in a single unified system all his own findings, as well as those of the astronomers and physicists who preceded him.

IS MAN A PART OF NATURE OR ABOVE NATURE?

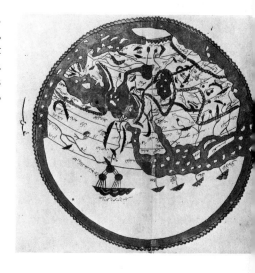

The universe is almost unimaginably immense. Galaxies billions of light years away have been found by modern astronomers. But most of the cosmic space is empty; stars, some of them enormous in size compared with earthly scales and exceedingly hot, are as dust in the cold cosmic immensity. Man feels lonely and lost amidst these cold wastes. More than three centuries ago, Pascal exclaimed: 'The eternal silence of these infinite spaces frightens me.' Man has found out that he inhabits merely a smallish planet revolving around a sun which is unremarkable either in size or in other qualities among countless stars. It is hard to believe that the universe has been created for man or because of man.

Descartes, a senior contemporary of Pascal, made a gallant effort to restore man's primacy. As a scientist, Descartes was far ahead of his times. He was what later became known among biologists as a mechanist, rather than a vitalist. That is, he believed that living bodies, including human bodies, were machines which could be studied and understood in terms of physical

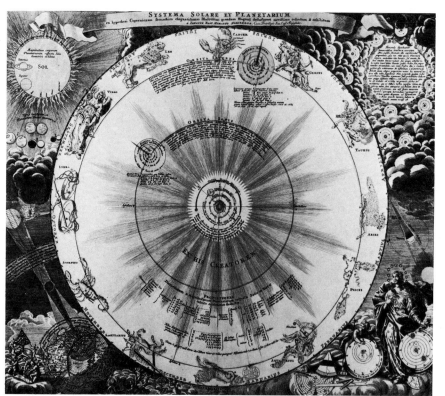

Ills. 185–187. Above: an Arabic map of the known world dating from AD 1154 which shows the fixed cosmos of antiquity and medieval times. This depiction is similar to that shown in Ptolemy's map of the world (*Ill. 12*), having no doubt been preserved by Arab astronomers—as was much classical science—throughout the tribulation of the Dark Ages. Nicolaus Copernicus (1473–1543) replaced this geocentric view of the universe and correctly inferred that the sun was the centre of our planetry system and that the universe was much vaster than had previously been imagined (left). The view (opposite) of the Horse-Head Nebula, south of Orion, shows that although modern optical and radio astronomy has pushed back the known limits of the universe—Pascal's 'infinite spaces'—Copernicus' basic theory is still valid.

principles that are known also in the non-living nature. He likened the body of a living man to a watch in good mechanical order, and a dead body to the same watch 'when it is broken and when the principle of its movement ceases to act'. Nevertheless, Descartes made a clear-cut distinction between humans and other living creatures. Animals are mechanical automata; they lack consciousness and are incapable of feeling. Animals have no souls, while humans have souls, each individual his own. Although man's body is a living machine, like the body of an animal or a plant, the possession of the soul sets man apart and above the rest of nature.

Ill. 190

The Cartesian body–soul dualism was unconvincing. It is a piece of religious folklore grafted on to an incompatible philosophical system. This dualism does however persist as a popular creed. As a scientific theory, it was abandoned even by those who accepted other principles of Cartesianism, because it becomes bogged down in insoluble difficulties. Indeed, let us suppose that the soul is an incorporeal agent; how does it exercise its influences on the body without itself becoming a part of the material world? Descartes believed that the soul resides in the pineal gland, and while it cannot influence the amount, it can alter the direction of 'vital spirits'. But where does one find the 'vital spirits'?

During the eighteenth, nineteenth, and early twentieth centuries there were lively controversies between mechanists and vitalists among biologists. These debates had in fact not much relevance to the problem of man's special position in the order of things, because vitalists claimed that all living beings, not man alone, are carriers of the immaterial 'vital force' or 'entelechy'. Furthermore, mechanism has been overwhelmingly victorious. So much so, that the few surviving vitalists who are professional biologists try to disguise their views under a variety of labels, in order to avoid being called vitalists.

The ascendancy of mechanism over vitalism is pretty generally known, not only by biologists but also by laymen. However, the real reasons for this ascendancy are not always understood. Biological research has made it possible to account for a multitude of biological phenomena by analysing them in terms of their chemical and physical components. Nowhere has it been necessary to invoke occult or special vital or psychical forces. Of course, this does not mean that all biological processes and phenomena have been so accounted for, and reduced to specialized patterns of physical and chemical forces and reactions. There will of necessity always remain a possibility that something, somewhere, is a product of some kind of vital force. Biology has elucidated many things, but many more remain to be explained. One example will suffice.

Much is known about the physiology of the nervous system, including the brain; yet nobody knows how it is that the brain generates thought and 'mind'. Sherrington was a great physiologist and no vitalist. Yet he wrote:[43] 'A radical distinction has therefore arisen between life and mind. The former is an affair of chemistry and physics; the latter escapes chemistry and physics.' Monod is assuredly no vitalist either, but he asserts:[35] 'There lies the frontier, still almost as impassable for us as it was for Descartes. What doubt can there be of the presence of the spirit within us? To give up the illusion that sees in it an immaterial "substance" is not to deny the existence of the soul, but on the contrary to begin to recognize the complexity, the richness, the unfathomable profundity of the genetic and cultural heritage and of the per-

Ill. 188. Opposite: the Bible story of the Creation and 'Fall' of man was often illustrated and these depictions not only helped to form medieval man's concept of the underlying religious message, but also supplied a 'scientific' explanation to the mystery of man's origins which was to be accepted until Darwin. Here, in a miniature from an early illustrated version of the *Septuagint*, the accusing finger of God points to Adam and Eve, who guiltily hide their nakedness behind stylized bushes.

sonal experience, conscious or otherwise, which together constitute this being of ours: the unique and irrefutable witness to itself.' I fully agree with Monod's argument, although I would hesitate using the word 'soul' in this context. 'Mind' or 'self-awareness' are better designations. The conclusion of paramount importance to which evolutionary biology leads us is that man is a part of nature and at the same time above nature.

DARWIN VERSUS COPERNICUS

An article which I published several years ago was entitled 'Darwin versus Copernicus'.[12] The idea epitomized in this title was rather simple but perhaps novel. In one sense, Darwin extended and completed the work of Copernicus. Copernicus deprived man's abode, the earth, of the status of the centre of the universe. Darwin showed that mankind is not a product of special creation separate from the rest of the living world. Our ancestors were animals, not humans. Mankind is a biological species.

According to some authors,[34] Darwin completed the 'breaking of man's image'. Yet in another sense, Darwin contravened Copernicus. Modern evolutionism, which stems from Darwin, finds that mankind and the universe are not static, not elementally unchanging, but on the contrary are involved in a process of perpetual change. There were in the past heavens and earth different from those we see today, and there will be still different ones in the future. In the pithy phrase of Teilhard de Chardin,[48] the cosmos has become cosmogenesis. In a past, geologically not remote, mankind did not exist; it evolved gradually, and it continues to evolve. The enterprise of creation was not completed some six thousand years ago, it is going on now. Man has discovered that mankind, the Earth, and the universe are evolving. This discovery is the source of hope that did not exist in a world which was believed to be unchanging. Knowledge and understanding of change may give to man a power not simply to witness the changes that are going on, but also to participate actively in, and perhaps to direct, the changes.

Darwin consistently refrained from discussions of philosophical implications of evolutionism (except, to some extent, in private correspondence). In point of fact, he rarely mentioned the word 'evolution'. This might have been to avoid possible ambiguity, because 'evolution' had been used to describe the development of a fertilized egg cell giving rise to an embryo, an infant, a juvenile, and an adult organism. To 'evolve' meant to unfold or unroll, to spread forth to view something that had been present but invisible earlier. Indeed, this was the way the biological evolution was represented, or rather misrepresented, by the exponents of theories of finalism, autogenesis, orthogenesis, nomogenesis, etc.

These theories arose after Darwin, and they opposed Darwin's conception that natural selection is the main propellant of evolutionary change. They did not restore the Cartesian dualism and the primacy of the human species. Some believers in finalism posited that mankind, as well as other now-existing forms of life, were predetermined to arise, and that they were potentially present in the primordial life. It merely took time to have their ancient disguises removed and their presence finally revealed. Very often, thought not invariably, partisans of autogenetic theories were also vitalists and believers in evolution being guided by occult, preternatural agencies, or even by God himself. The few who were not vitalists had to postulate yet

Ill. 189. Opposite: this imaginative photo-montage by John Garrett symbolizes the eternal quest of the human spirit.

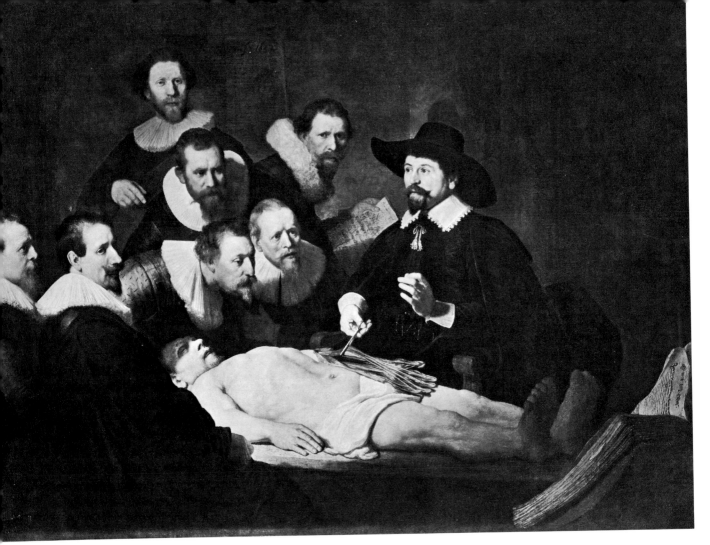

undiscovered forces, because the known genetic phenomena could not easily be imagined to add up to autogenesis. Autogenesis is now a minority creed, surviving mainly in France and some other European countries.[20]

Darwin and most of his successors have, very sensibly, postponed philosophizing for the sake of more pressing tasks. These were simply to gather evidence that evolution did in fact occur in the history of the earth, and to clarify the causes, principally natural selection and mutation, that bring evolution about. It was particularly important to verify that mankind is a product of evolutionary development, because this discovery was, and in some places still is, opposed with greatest vehemence. More and more evidence of the evolutionary origin of man continues, of course, to be generated by current research. However, by the beginning of the present century, and even earlier, this evidence became so overwhelming that further additions are exciting only if they are important for their own sakes, not only attesting the reality of evolution as a historical happening. It would surely be futile to try to convince the remaining doubters, who are either uninformed or bigots to whom no evidence is meaningful. The inspiring task is at present perhaps the reverse of what it used to be: while the classics of evolutionism endeavoured to show that mankind is like all other biological

Ill. 190. Rembrandt, a contemporary of Descartes, illustrates in the famous picture, *Dr. Nicholaes Tulp Demonstrating the Anatomy of the Arm,* the quest for anatomical and scientific knowledge which was beginning to grow in the seventeenth century, sparked off by Descartes' view of man as a machine.

species, at present we discover that in many significant ways mankind is unique and unlike anything else. The uniqueness of man surely did not escape the perception of our predecessors: it was rather de-emphasized in favour of what is in common between man and other organisms.

MAN AS A BIOLOGICAL SPECIES

It is not my intention to review here the evidence of mankind's ties with the rest of the living world. It will suffice to mention only some of the historical benchmarks (see also Greene[21] and Dobzhansky[9]). The evidence is, roughly, of two kinds. First, man's body structures, functions, and composition resemble those of other organisms, and the similarities are greater to some than to other beings now living. Secondly, the evidence of fossils permits us to trace, in more and more detail, the path of the evolutionary development which has culminated at our time level in mankind.

That human bodies have parts corresponding to those of animals, particularly mammals, is obvious to even casual observers. In 1555, Belon noted the corresponding, or as we would now say homologous, bones in the skeletons of men and a bird. In 1699, Tyson dissected an infant chimpanzee, and noted forty-eight characters in which it resembled man more than monkeys, and only thirty-four characters of the opposite kind. In several editions of *Systema Naturae* (1735–66), Linnaeus placed man squarely in the mammalian order Anthropomorpha (primates), together with apes. In his *Man's Place in Nature* (1863), T. H. Huxley reviewed all the pertinent evidence, and showed that man resembles apes in bodily structures more than any other animal. Evolution from common ancestors was the most probable explanation of the resemblances. This agreed with Darwin's view, implicit but not spelt out in *The Origin of Species*.[6] It was made explicit in *The Descent of Man*.[7] Haeckel in 1874 stressed the similarity of human embryos to those

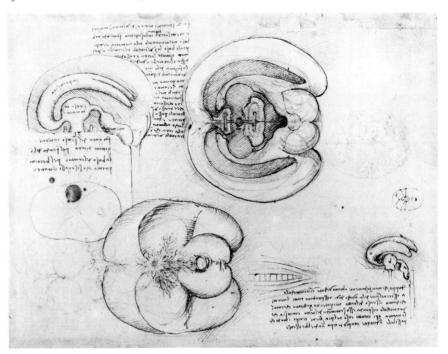

Ill. 191. Leonardo da Vinci, a hundred years before Rembrandt, had already become fascinated not only by man's anatomy, but also by the functioning of the brain, as these drawings from one of his anatomical sketchbooks show.

of other mammals, and in fact to those of other vertebrates, as corroboration of the common ancestry.

Recent advances of molecular biology make possible precise measurements of the evolutionary similarities and distances between various forms of life.[13] Twenty kinds of amino acids are common constituents of proteins. Protein molecules are chains of amino acids following each other in a fixed order; the linear sequences of the amino acids have been determined for some proteins in different animal and plant species. For example, the alpha chain of haemoglobin has 141 amino acids, and these 141 have identical arrangements in man and in chimpanzee. Human alpha differs in 1 amino acid from that of gorilla, in 17 from that of cattle, in 25 from rabbit, and in 71 from a fish (carp). Equally impressive are the data on another protein—cytochrome c. The cytochrome c molecule is a chain of 104 amino acids in man and other vertebrate animals. The cytochromes of man and monkey (macaca) differ in a single amino acid substitution. The mutational distance (the number of mutational changes needed to transform one cytochrome c into another) between man and dog or pig is 13, man and birds 16–18, fish (tuna) 31, man and molds or yeasts 53–56. Unless man and yeast had a common ancestor in some remote past, there is no reason why the cytochromes c should have any recognizable similarity in the amino acid sequences. Nor is there any reason why the cytochromes of man and monkey should resemble each other more than either resembles the cytochrome of yeast, unless the common ancestor of man and monkey is much more recent than that of man and yeast.

Ill. 184

OUR HUMAN AND PREHUMAN ANCESTORS

Some people refuse to accept similarity of body structure or composition as testimony of common descent. P. M. Gosse, one of the early opponents of evolutionary ideas, argued that God has seen fit deliberately to arrange things as if evolution was happening, although in truth there was no evolution. Arguments of this kind are repeated from time to time by anti-evolutionists now living. One stands in amazement before a mentality capable of taking seriously such arguments—is it not obvious that they are blasphemies, accusing God of absurd deceptions? Anyway, already T. H. Huxley recognized the need for palaeontological evidence to establish the stages which the process of hominization has actually gone through. From then to the present, palaeontologists have been looking for something called, at least in popular accounts, the 'missing link', supposed to have been a creature midway in its characters between ape and man. The rash but perspicacious Haeckel a century ago ventured even to name the hypothetical 'missing link' before it was discovered—*Pithecanthropus alalus*, i.e. ape-man without speech. His audacity was vindicated: not one but a whole series of links were gradually brought to light.[5,28,49,50]

Bones of Neanderthal man were discovered in 1857, i.e., before Darwin's *The Origin of Species*[6] was published. The disputes which this find aroused within the scientific establishment of that time seem amusing in retrospect. The opinions ranged from that the Neanderthal bones belonged to an extinct ape-like species, to that they were remains of a pathological specimen of modern man, or perhaps of a Napoleonic soldier with water on the brain who died on the retreat from Moscow in 1812! The truth, according to the present views, is about midway between the above extremes: the sub-species *Homo sapiens neanderthalensis* inhabited a territory extending from Western Europe to Palestine and Central Asia, from roughly 100,000 to 40,000 years ago. A Dutch physician, Eugene Dubois, had a hunch that the 'missing link' would be discovered in Asia, went to the former Dutch East Indies, and in 1891 found in Java a skull cap, a femur, and some teeth, the whole fitting his idea of what the 'missing link' should have been. The creature had a cranial capacity of about 850–900 cc, i.e., between that of the chimpanzee and modern man. It must have walked erect, as shown by its femur bone. It was accordingly named *Pithecanthropus erectus*, but is at present usually considered to belong to the genus *Homo* and is called *Homo erectus*. A subspecies *Homo erectus pekinensis* lived about 300,000 to 400,000 years ago in China, had an average skull capacity of about 1000 cc, and was the first known user of fire. Other varieties of *Homo erectus* were found in Africa, in Europe, and at different time-levels in Java.

The most remarkable 'links', no longer missing, between man and his ape-like ancestors have been found in South Africa since 1924 by Raymond Dart, Robert Broom, and their successors, and in East Africa since 1959 by Louis Leakey and members of his family. The discoverers have invented a plethora of generic and specific names for their finds, but all or most of these can be placed in two fossil species, *Australopithecus africanus* and *Australopithecus robustus*. They belong to the subfamily Australopithecinae, which some authorities classify in the human family, Hominidae, and others in the ape family, Pongidae. Their cranial capacities were usually below 700 cc, and even below 650 cc. This is within the range of cranial capacities in the now-living apes, but these apes are much larger animals than the australopithecines were, so that the brains of the latter must have been relatively larger. The pelvis bones show that the australopithecines had achieved erect posture and bipedal locomotion, although perhaps not to the extent that man has. Their age was at first underestimated, because the localities in South Africa where they were discovered do not lend themselves easily for such estimates. The date of 1·75 million years for some East African finds seemed surprisingly high. Yet more recently reported finds are older still—2·6 million, or even more, years.

Perhaps the most exciting find of all, made by the Leakeys in 1961 and critically studied by P. V. Tobias, is a close relative of *Australopithecus africanus*, but with a somewhat greater cranial capacity. Its remains were found associated with very primitive but unmistakably deliberately made stone tools. It was given the name *Homo habilis*, the skilful man. Some authors prefer to call it the subspecies *habilis* of *Australopithecus africanus*. Placing it in the genus *Homo* suggests that it has crossed the threshold between animality and humanity; leaving it in *Australopithecus* implies that it has not yet crossed this threshold. The 'threshold' is however a metaphor; there is no dividing line that would not be at least to some extent arbitrary. What is important is that the gradual evolution of man from prehuman ancestors is substantiated by reliable evidence.

MANKIND—AN UNUSUAL ANIMAL SPECIES

P. Teilhard de Chardin and G. G. Simpson are two scientists who seldom agree on anything. But they do agree that the species mankind is a quite extraordinary product of the evolutionary process. Teilhard wrote:[48] 'Man, as science is able to reconstruct him today, is an animal like the others. . . . Yet, to judge by the biological results of his advent, is he not in reality something altogether different?' And Simpson:[45] man 'is another species of animal, but not just another animal. He is unique in peculiar and extraordinarily significant ways.' What are these significant ways in which man is something altogether different? They pertain mainly not to body structures but to behaviour and mental capacities. Biological classification always was and still is based very largely on comparative morphology, the study of conformation of body parts. This is all that is directly available for examination in fossil remains of our ancestors and relatives. Behaviour is not fossilized, it can only in some instances be inferred from the environmental conditions in which a fossil is found, or from the presence of artifacts.

On morphological grounds, mankind is classified as the sole surviving species of the family Hominidae. Nobody would erect a separate zoological order, let alone a class or a phylum, for this single species, although it might well be different if the classification was based on psychological characteristics.

Of course, most bones, muscles, and internal organs of the human body are recognizably different from those of a chimpanzee or a gorilla. Simpson[45,46] gives the following list of 'the most striking anatomical features of *Homo sapiens*': (1) cranium globular, brain large relative to body size; (2) *Ill. 193* face short, almost vertical; (3) orbits directed anteriorly, fields widely overlapping; (4) dentition modally 2.1.2.3/2.1.2.3, brachydont, series closed, canines small, cheek teeth bunodont, molars subquadrate; (5) posture erect, spinal column sigmoid; (6) arms shorter than legs; (7) thumbs well- *Ill. 192* developed, opposable; (8) ilium short craniocaudally, expanded dorsoventrally; (9) foot plantigrade at rest or in slow locomotion; (10) great toe not opposable, modally nearly or quite as long as the second toe.

At least half of the traits listed, and many unlisted ones (especially those of musculature), are functionally related to man's erect posture. Man is one

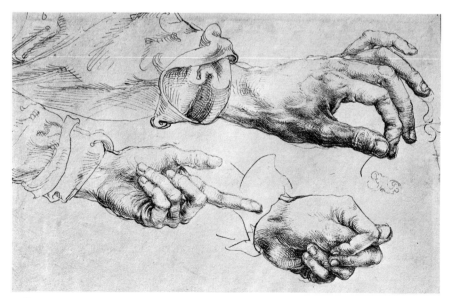

Ills. 192, 193. Though in many ways similar, the human body is recognizably different from that of other primates. The most striking physical features of *Homo sapiens* are shown opposite, in the measured anatomical drawing by Albrecht Dürer, who also drew the studies of hands (left) which demonstrate one of the features unique to man: his fully opposable thumb.

(1) Globular cranium
(2) Short, almost vertical face
(3) Forward-facing eyes with overlapping field of vision

(5) Erect posture

(6) Arms shorter than legs
(7) Opposite: opposed thumbs

(9) Foot plantigrade
(10) Non-opposable great toe.

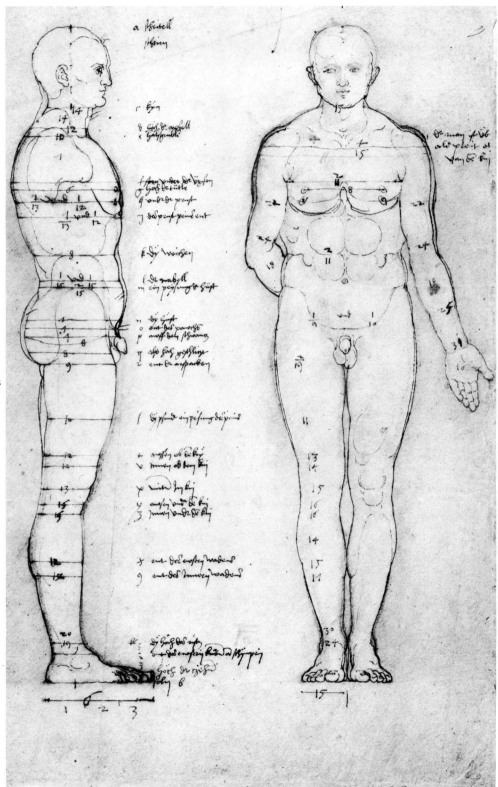

of the few, but not the only, species of mammals which normally use their posterior but not their anterior extremities in locomotion. Other traits are connected with the enlargement of the brain. Even so, there is nothing strikingly unique and unprecedented about human morphology. Nor is the human body in any meaningful sense the most advanced or nearest to perfection among animals. Human vision is good but inferior to that of, for example, some birds of prey. Our olfactory sense is far inferior to that of dogs and many other animals. Hearing is good but not exceptionally so. It is among mental traits that human biological uniqueness is most strikingly manifested.

According to Susanne Langer,[30] 'the trait that sets human mentality apart from every other is its preoccupation with symbols . . . all human activity is based on the appreciation and use of symbols. Language, religion, mathematics, all learning, all science and superstition, even right and wrong, are products of symbolic expression rather than direct experience.' A symbol is an object or an act whose meaning is socially agreed upon, or, according to Leslie White,[52] is 'bestowed upon it by those who use it'. Human languages are symbolic, in contrast to so-called animal 'languages', which consist of signs rather than symbols. To quote Langer again: 'A sign is anything that announces the existence or the imminence of some event, the presence of a thing or a person, or a change in a state of affairs.' Communication by signs is widespread; animals as well as men use signs, but only humans create and use symbols, and communicate by means of symbolic languages.

ANIMAL ANTECEDENTS OF UNIQUE HUMAN CHARACTERS

A novel biological system can be constructed in evolution only if the raw materials for its construction are available for natural selection to work with. At least some of the genetic building blocks from which the novel system is to be compounded must be older than the system itself. For this reason, it is a challenging problem for evolutionists to discover whether or not some rudiments of symbolic communication and of other uniquely human qualities are found on the animal level. However extraordinary and unusual these unique qualities may be, their rudiments may occur in animals other than man. The capacities to use and to make tools, or to acquire and utilize human language, do not cease to be uniquely human characteristics if we find them foreshadowed in some other species. Two types of research may be expected to yield relevant information. First, such information may come from observations on the behaviour of animals, especially those most closely related to man, in their natural habitats. Secondly, one may try to induce some man-like behaviours in experiments on animals in captivity. Both approaches have been successfully utilized.

By far the most productive in recent years have been the observations of Jane van Lawick-Goodall and her collaborators on wild chimpanzees [31,32,33] (see Chapter 6), of Schaller on gorillas, of DeVore and others on baboons and other primates.[8,23] Man is dependent for his survival on using and making tools. Instances of tool-*using* have occasionally been recorded in a variety of animals, down to some invertebrates. Tool-*making*, for use in even the near future, is very rare. Fascinating observations, however, have been made of termite 'fishing' by chimpanzees, using a twig stripped of its leaves.

Ill. 149
Ill. 150

Ill. 143

Ill. 194. Symbolic communication exists not only between men and the higher animals but also in the 'dance' of worker bees in the hive, during which they impart information to one another about the direction and distance of a source of food. In contrast to man, the behaviour of the 'dancing' bees is genetically determined.

Sticks are often used to enlarge holes or openings where food can be found, or to investigate objects which the animal is afraid to touch. Bunches of leaves are used as sponges to get water which the chimpanzee cannot reach with its lips, and to wipe off dirt from its body. In some localities, but not in others, chimpanzees have been seen using stones to open hard-shelled fruits. This tool-using and tool-making develop into a kind of rudimentary 'culture', since the juveniles attentively watch and imitate the behaviour of their elders.

Van Lawick-Goodall and others have shown that chimpanzees are by no means exclusively vegetarian (see pp. 151–3). Not only do they engage in hunting other animals to secure meat, but they may do so in groups of several individuals that co-operate with each other. Moreover, when the hunt

Ill. 151

is successful, the hunters share the meat among themselves. This is most suggestive because many authors have quite plausibly speculated that co-operation in hunting played an important role in human evolution, favouring the development of communication by symbolic language and other human characteristics.[8,51] The discovery of hunting in chimpanzees makes it tempting to suppose that this form of behaviour may have already been present in the common ancestors of man and chimpanzee, which would make it ten or more million years old.

It is obviously important to discover whether at least some rudiments of symbolic communication exist on the animal level. Oddly enough, it was not among higher vertebrates but among insects that what has the appearance of

Ill. 194

symbolic communication was first discovered. The classical experiments of

von Frisch[17] have demonstrated that a honey bee imparts to her hivemates the information about the direction and distance of a source of food by means of a 'dance' on the surface of the comb. The meaning of these apparently symbolic dances is, however, genetically fixed, rather than learnt or agreed upon.

Several investigators have attempted to induce learnt symbolic behaviour in higher primates by making them acquire rudiments of human symbolic language. Chimpanzees and some other higher primates have large 'vocabularies' of calls, inborn rather than acquired entirely by imitation or teaching, the purport of which is readily understood by individuals other than the one that emits the call. Nevertheless, these animals have proved to be remarkably refractory in experiments that attempted to teach them human auditory language. Thus Hayes and Hayes[24,25] had little success, despite much effort, in teaching a chimpanzee child to speak. Gardner and Gardner[19] accomplished much more with another individual by teaching the chimpanzee not the ordinary auditory, but rather a sign language, consisting of gestures, and originally developed for communication between deaf-mute persons. Premack[38] was even more successful. He utilized as symbols figures cut from coloured plastics, provided with small iron plates by means of which they could be attached to a magnetized board. A chimpanzee first learnt symbols that stood for objects, then symbols for 'yes' and 'no', for 'similar' and 'different', colours and shapes. With this 'vocabulary' of symbols, the ape was able to understand not only words but also what amounted to simple sentences.

It is evident from these experiments that some animals other than man can, under human guidance, master the use of symbols. This detracts nothing

Ills. 195–197. Teaching symbolic communication to primates has proved extremely difficult. However, although teaching chimpanzees to speak has met with little success, by using Amerslan—American sign language—practised by the deaf and dumb, the Gardners taught Washoe, a female chimpanzee, a wide repertoire of signs which she would use,

Ills. 195–197

not only to ask for food, drink, etc., but also to construct sentences—as above, reading from left to right: you (give) drink (to) me. Washoe also used this language to express abstract concepts such as 'nice', 'stupid' and 'mine'. Dr. Roger Fouts, seen here with Washoe, continued and developed the Gardners' experiments.

from the significance of symbolic behaviour as a paramount human adaptive trait. But it does raise a problem. We have found out that building blocks from which the genetic basis of language and other symbolic behaviour was constructed in human evolution was already present on the prehuman level. What advantage could animals lacking speech derive from the possession of an ability to be taught the use of symbols? To be sure, an ape's ability to master symbolization is small compared with that of any non-pathological human child. Even so, it seems unlikely that natural selection would have promoted this ability, or that it could have become established if it contributed nothing to the adaptedness of its possessors. More ethological studies on primates, particularly the great apes, will perhaps solve this puzzle.

The beginning of communication by symbolic languages in mankind cannot be dated, even approximately. The fact that all human populations without exception have languages suggests but does not prove the great antiquity of this form of communication. Holloway made, in 1969, a brave attempt to connect language with tool-making.[27] According to him, natural selection 'favoured the cognitive structures dependent on brain organization and social structure which resulted in both language and tool-making'. *Ill. 199* Holloway does not explicitly say whether making primitive 'pebble' tools, like those found with the remains of the australopithecines, indicates the existence of these cognitive structures and 'imposition of arbitrary form upon the environment'. The rudiments of tool-making discovered among chimpanzees presumably do not qualify as an indication of these cognitive structures. By contrast, some anthropologists, on grounds that do not seem to me to be particularly convincing, would make the appearance of language

coeval with the species *Homo sapiens*. This would make the most ancient Prometheus, *Homo erectus pekinensis*, without language.

Ills. 84, 198

There is no doubt that communication by symbolic language was promoted by natural selection. The cognitive structures on which both language and tool-making depend are the foundations of adaptation to and control of the environment by culture. To be very concise: while all organisms adapt to their environments by changing their genes, man alone adapts mainly, though not exclusively, by creating new environments to fit his genes. In so doing man has often abused the environments which he inhabits, and the abuse becomes more and more serious as the numbers of people increase. Man must be more circumspect than he habitually has been in dealing with the overcrowded little planet that is his abode. Yet it can be affirmed without hesitation that it was the genetically based cultural development which lifted the human species up to its present estate, sometimes described with no little arrogance as 'The Lord of Creation'. Bidney[3] has stated the unique features of human psyche admirably as follows: 'Man is a self-reflecting animal in that he alone has the ability to objectify himself, to stand apart from himself, as it were, and to consider the kind of being he is and what it is that he wants to do and to become. Other animals may be conscious of their affects and the objects perceived; man alone is capable of reflection, of self-consciousness, of thinking of himself as an object.'

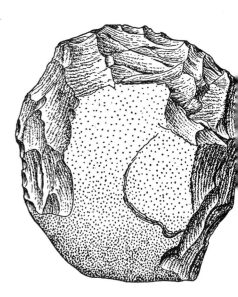

EVOLUTIONARY TRANSCENDENCE

Is it necessary to stress so much the biological uniqueness of the human species? To some people this seems almost a rejection and betrayal of Darwin. But it is nothing of the sort. Except for an insignificant minority of obscurantists and fanatics, everybody acknowledges that mankind has emerged from the animal kingdom. This is no longer news. What needs emphasis now is that evolution does not proceed everywhere in the same manner. In particular, in the line of ascent leading to the human species, the evolutionary process has transcended itself.

Organisms do not evolve for the sake of evolving, they evolve to improve their adaptedness to the old, and to capture new ways of living. Most often evolution is cladogenesis, the splitting and re-splitting of the ancestral stem into numerous branches and twigs of derived species. Less frequently, evolutionary lineages invade new environments and adopt novel ways of life. This is anagenesis and progressive evolution. In more or less constant environments, the occupants of some unchanging ecological niches may evolve little or not at all. The results are the so-called 'living fossils'.[39,44] The least frequent kind of evolutionary change is transcendence, which means going beyond the limits of, or surpassing the ordinary and previously utilized possibilities of a system.

We know two major transcendences in the evolutionary history of the universe. The first was the origin of life from inanimate matter. The second was the origin of man from his animal ancestors. I am well aware that the word 'transcendence' is jarring to the ears of many of my scientific colleagues. Perhaps 'emergence' would be a preferable word. Anyway 'transcendence' or 'emergence' does not imply intervention of any agent which has not acted in evolutionary changes of lesser magnitude.

Transcendences do however usher in new realms of phenomena and of natural laws that simply had not existed before the transcendences occurred. Thus, phenomena of adaptedness and adaptation are basic on the biological level, but it would make no sense to talk of adaptedness of mountains or of stars. There were no Mendel's laws before the appearance of life and of sexual reproduction. Individuality, natural selection, populations, ecological communities, etc., are biological, not chemical or physical phenomena. This means that they are patterns or states or events which occur only in the biological dimension, not that they somehow breach the laws of physics. Not even a vitalist would claim that Mendel's laws are manifestations of entelechy or vital force.

Good and evil, altruism and egotism, the sacred and the profane, mythology, poetry, mysticism, religion—all of these exist only in the human dimension, even though we may endeavour to find their germs on the animal level. They are not biological phenomena without, certainly, infringing on or breaking the laws of biology. It may be helpful at this point to be reminded that, in addition to the two major transcendences, there were some lesser ones in the evolutionary history. One example is the origin of eucaryotic from procaryotic cells, possibly by capture and incorporation of symbionts. Another example is the origin of land life, or of the maintenance of constant body temperature.

It is in the realm of psychic, rather than of body functioning, that man differs radically from other organisms. This statement should not be mis-

Ills. 198, 199. Peking man was one of the earliest users of fire and his primitive tools have been found in the caves at Chou-k'ou-tien with the remains of his camp fires (opposite, below). The reconstruction by Maurice Wilson (opposite) shows him at work in the mouth of a cave, making some of the stone tools with which he is associated.

understood as a bow in the direction of Cartesian dualism, any more than the unsolved (but hopefully not insoluble) mysteries of the human mind and 'soul', pointed out by Sherrington[43] and Monod[35] (p. 192), revived such dualisms. It is a fact that my 'mind' or 'soul' or 'spirit' is *for myself* the prime reality, so evident that it needs no proof. This is what Descartes was unable to doubt: *Cogito ergo sum*.

This reality is, however, troublesome when approached scientifically. I cannot doubt the reality of my mind, because my conscious self is my mind. The validity of the evidence weakens at once when it comes to the minds of other people. I cannot get inside anyone else's mind. To what extent do the perceptions and experiences of my friends and neighbours resemble my own? A simple consideration shows that the difficulty is far from fictitious. Between five and nine per cent of males in Caucasian populations are red-green colour-blind. To a person who is not colour-blind, red and green objects look unmistakably different. But how can one discover how they appear to a colour-blind person? Does he see them both as I see red, or green, or perhaps as grey objects? For that matter, do all persons who are not colour-blind see red colour as I do?

To avoid uncomfortable solipsism, I assume that other people have minds approximately like my own mind. The evidence supporting this assumption is that most people act most of the time as I do under similar circumstances, and I know that my mind is active when I so act. However, as soon as one leaves the confines of the species mankind, the evidence of the possible existence of something like mind becomes so tenuous as to be worthless scientifically.

There are respectable scientists who are convinced that animals, at least some mammals and birds, have rudimentary but functioning minds. Others, such as Birch[4] and Rensch,[40] go so far as to ascribe some 'protopsychic' functions not only to all living beings but to molecules, atoms, and electrons. They are followers of the so-called process philosophers, of whom A. N. Whitehead[53] was the most eminent pioneer. At the opposite extreme, there is the influential school of neo-behaviourists, who blandly refuse to deal with 'minds'. Animals are 'black boxes'; instead of trying to find out what goes on inside a 'black box', one should occupy oneself more profitably by recording what comes out of it as observable behaviour. This is all to the good, but the problem of the mind refuses to vanish because some people deny its existence.

SELF-AWARENESS AND DEATH-AWARENESS

Mind, consciousness, or self-awareness are alternative terms that have much the same referents—the existential actuality familiar to everyone through introspection, but elusive when it comes to its presence in others. Nothing could ever make me doubt that I am aware of myself (except when asleep or under narcosis), but I infer that other humans also possess self-awareness only by analogy with my own self-awareness. We cannot either affirm or deny that some beginnings of self-awareness may be present in species other than mankind. In any case, the difference between mankind and other animal species is, in this respect at least, quantitatively so great that it amounts to a qualitative difference.*

Fromm[18] has stated the distinction between man and animal as follows:

Ill. 200. Opposite: this allegorical painting by Hans Baldung, depicting vanity, typifies man's awareness of himself, of his past and of his future which is death.

* A minutiously careful but rather casuistical analysis of the Cartesian problem of whether man 'differs in in kind' or 'only in degree' from the rest of nature can be found in Adler.[1]

Man has intelligence, like other animals, which permits him to use thought processes for the attainment of immediate, practical aims; but man has another mental quality which the animal lacks. He is aware of himself, of his past, and of his future which is death; of his smallness and powerlessness; he is aware of others as others—as friends, enemies, or as strangers. Man transcends all other life because he is, for the first time, life aware of itself. Man is in nature, subject to its dictates and accidents, yet he transcends nature because he lacks the unawareness which makes the animal a part of nature—as one with it.

Man possesses self-awareness and death-awareness. These are twin characteristics unique to man, even if their rudiments were found in other species of animals. All animals sooner or later die, but there is no evidence that individuals of any species other than man know that they will inevitably die. Van Lawick-Goodall and Hamburg[33] describe a chimpanzee juvenile who exhibited a behaviour clearly indicative of 'grief' when his mother died. This certainly does not prove (and the authors do not claim that it does) that the grieving juvenile realized that some day he will suffer a fate like his mother. Some authors believe that certain animals (elephants) withdraw to places analogous to cemeteries when they feel that death is approaching; this story is unconfirmed and undoubtedly erroneous. One can have more assurance that animals lack death-awareness than that they lack self-awareness. Indeed, while self-awareness is really known only from introspection, death-awareness manifests itself in objectively observable behaviour of which it is the source. This is ceremonial burial of the dead.

Burying the dead is universal in man, and is absent in species other than man. Burial rites are quite different in different cultures: interment, cremation, exposure to carrion-eating birds or beasts. One thing that does not

Ills. 201, 202. In this cave (above left) in Iraq, Professor Ralph Solecki discovered several Neanderthal burials. One of these was found to have been covered with adornments, one of the earliest proofs of man's preoccupation with death (above right).

happen, or at least is exceedingly rare in man, is discarding the cadaver and paying no more attention to it than to other rubbish. Yet this is exactly what happens in non-human species. Ceremonial burial is no recent invention. It is at least as ancient as Neanderthal man, and possibly Peking man. For perhaps as long as half a million years humans have been preoccupied with death. They were concerned about the death of others because they knew that theirs would be a similar fate. Animals do not know and, except for mothers and their young, are indifferent to the death of other individuals.

The diversity and universality of burial rites in man, and their absence in animals, must be understood in the context of evolution. Evolution is in the main utilitarian. How did it come about that self-awareness and death-awareness became universal characteristics of the human species? Did they enhance the adaptedness or Darwinian fitness of their possessors, so that natural selection promoted and established them in mankind? The pragmatic significance of self-awareness may well be quite different from that of death-awareness. The adaptive role of self-awareness is sufficiently obvious, no matter how elusive it is because of being known mainly from introspection. As pointed out by many anthropologists and psychologists (e.g., Hallowell[22] and Hofer and Altner[26]), self-awareness is the foundation of the specifically human personality structure. It is the basis of human social organization which has made man 'The Lord of Creation'. It is an integral part of the adaptive complex which includes also the use of symbols, symbolic language, and hence acquisition and transmission of culture. Self-awareness is manifested in childhood and persists till death. Only some rare individuals with gross pathologies appear to lack self-awareness. They are eliminated by early death, unless maintained by artificial means. There is no way of establishing even an approximate date when self-awareness appeared in human evolution. It is hard to conceive that it could have appeared suddenly and fully developed. It must have developed gradually, as natural selection advanced it further and further, thus raising the adaptedness and adaptability of its possessors. The evolutionary transcendence of which mankind is the fruit arose hand in hand with the development of self-awareness.

The adaptive function of death-awareness is not as clear as that of self-awareness. What conceivable advantage could our remote ancestors at the dawn of the hominization process have derived from knowing that they would inevitably die some day? To be sure, this knowledge became useful when parents began to make provision for the welfare of their offspring after their own demise. This would seem, however, to presuppose a rather advanced stage of cultural development. Parents must foresee the situation which will arise after their death, and make altruistic choices and arrangements for the benefit of their progeny. Many animals to which there is no reason to ascribe death-awareness achieve the same end by means of genetically programmed instinctual behaviour. Thus, many insects build and provision nests for their offspring whom they never see except as eggs which they deposit. Of course, the same adaptive function is frequently achieved by different means in different organisms. However, it is improbable that at its inception the death-awareness functioned for the benefit of the progeny.

I have suggested[11] that the death-awareness arose in man originally not because it was adaptively useful by itself, but because it was a by-product of the self-awareness which was adaptive. Genes determine not separate and

Ill. 203. Not only was prehistoric man aware of death, he was also very aware of himself. This tiny ivory head of a French Gravettian 'Venus' figurine shows clearly marked features and an elaborate hairdressing which is unusual in figurines of this period.

Ill. 204. The Victorian epoch was very aware of death and made much of it. This highly melodramatic print is typical of the almost theatrical performance which surrounded death and of which children from their earliest days were well aware.

independent 'characters', but rather developmental processes through which certain characteristics of the organism are brought into being. Genes that cause the development of several characters are termed pleiotropic, and most genes are pleiotropic to various extents. One need not make an utterly improbable assumption that there is a single gene 'for' self-awareness and death-awareness together. It is rather a complex of many genes. Death-awareness is hardly possible without self-awareness, but with the progress of the latter the anticipation and foreknowledge of death would almost necessarily arise. It is interesting that in human development death-awareness appears later than self-awareness. Children sometimes do not accept the idea of their own mortality until the age of eight or nine years or later, and this even after they witness a death of a member of their family. This fact need not be interpreted as an instance of the biogenetic 'law'. What it suggests is rather that death-awareness is a derivative of self-awareness.

THE FALL OR RISE OF MAN

A poetic rendering of man's evolutionary transcendence can be seen in the biblical narrative of eating the fruit of the tree of the knowledge of good and evil. Man ate the forbidden fruit. 'And the Lord God said, Behold, the man is become as one of us, to know good and evil: and now, lest he put forth *Ill. 205* his hand, and take also of the tree of life, and eat, and live forever: Therefore the Lord God sent him forth from the garden of Eden, to till the ground from whence he was taken.'

Our remote ancestors were lowly animals, not humans. Animals are as unaware of evil and sin as they are of good and virtue. A predator captures and devours his prey. It would be ludicrous to regard him cruel and wicked. In many species of animals the parents defend their young, often risking their lives in the process. It would be senseless to credit them with heroism or altruism. Their actions are not free choices, they are programmed in their genes. When a honeybee defending its hive stings a presumed enemy, it commits suicide; its stinger has reversed barbs, and while stinging it eviscerates itself. It would be absurd to esteem this act as identical with *Ill. 206* human self-sacrifice. Egotism, altruism, heroism can exist only where the actor who performs a deed is conscious of its ethical value, and can proceed or abstain from its performance. Presumably only man can do that.

At some point, or rather at some stage of its evolution, mankind became able to distinguish right from wrong, good from evil. Tradition calls this the Fall of Man. Man 'fell' when for the first time he became able to, and did, freely transgress an ethical precept. Yet by the same act man rose above the level of animality. The innocence of the animal state is due to the animal's inability to know that any of its acts are unethical or evil. Man can be evil because he knows. According to the biblical narrative, the temptation came from the serpent: 'For God doth know that the day ye eat thereof, then your eyes shall be opened, and ye shall be as gods, knowing good and evil.' In the second century AD, the Gnostic sect of Ophites (snake worshippers) venerated the serpent, the tempter of Eve, as a beneficent deity. What a bizarre idea! And yet the transition from the state when the concepts of good and evil were absent to a human person who has these concepts would seem to constitute an ascent to a higher level of being, rather than a downfall. Man has risen, not fallen.

Christian theology has elaborated the idea, inherited from Judaism and from philosophers of the Hellenistic Age, of a Golden Age at the beginning of the world, and of regressive evolution thereafter. This idea took the form of a belief in the Original Sin and the Fall of Man. Adam and Eve disobeyed God's commandment, and gained knowledge of good and evil. The disobedience was sin which corrupted the human nature. It corrupted not only the sinners themselves but also all their descendants, including the mankind now living. John Calvin went so far as to declare man to be totally depraved. Can this be reconciled with evolutionary thought? The doctrine of Original Sin was at the root of the conflict between the evolutionist world view of

Teilhard de Chardin[48] and his ecclesiastic superiors. If the biblical account is taken literally, it is hard to see how a sin of a remote ancestor can be 'inherited' by all his descendants. It is even harder to comprehend why the descendants deserve punishments for a sin they did not commit.

The evolution did not take a regressive step with the appearance of man. Quite the opposite—the entry of man on the world scene was a leap forward in the history of the cosmos. Sin, misdeed, crime, transgression can occur in a being who is aware of himself and of the nature of his acts. Man is such a being, but his ancestors were not. One can admire the guiltlessness of an animal, but one can hardly wish that man would relapse to the animal state and become blind to the distinction of good and evil.

WHAT ARE PEOPLE FOR?

The ability to ask such questions is part of the evidence of man's evolutionary transcendence. What do mice, or Drosophila flies, or pine trees think they are for? They have no such thoughts or questions. If you take a flight of fancy and imagine that some of them could make such an enquiry, the sole conceivable response would be that they live to be alive. Some people brush the question off—they live to have fun. But it is no good for four billion people inhabiting a small planet each to have fun by and for himself. This leads to misery for increasing majorities, and eventually for everybody. If we were innocent animals unaware of good and evil, there would be nothing that we could even try to do about it. Yet we can envisage possible futures, evaluate some of them as more good and desirable and others as less desirable or evil, and strive to bring the preferred ones to realization.

It is a sad and astonishing fact that many people, at least in the capitalist West, love to be told that science has, allegedly, demonstrated that man is

Ill. 205. Michelangelo's rendering of the Temptation and Fall is perhaps one of the most powerful depictions of the 'Tree of Life'.

biologically, and irrevocably, selfish, aggressive, and evil. To Morris[36] we are 'naked apes' (although *Homo sapiens* is not ape, and rarely goes naked). He rejects the idea '. . . that our intelligence can dominate all our basic biological urges. I submit that this is rubbish. Our raw animal nature will never permit it'. Ardrey[2] goes still further. We are 'Cain's Children. Man is a predator whose natural instinct is to kill with a weapon. The sudden addition of the enlarged brain to the equipment of an armed, already successful predatory animal created not only the human being but also the human predicament.' Skinner[47] sees man as 'beyond freedom and dignity', not because he lost them but because he never had and cannot possibly have them. The only way to manage mankind is by a conditioning programme, which combines the features of Aldous Huxley's 'Brave New World' anti-utopia, and Dostoevsky's formidable but sinister 'Grand Inquisitor'. The books of Morris, Ardrey and Skinner are among the bestsellers. People are prone to feel relieved if they can blame their vices on their biological 'nature' which is beyond their control.

Both the optimistic thesis that man is inherently good, and the pessimistic antithesis that he is fallen and depraved, antedate scientific biology and anthropology. Great religions, particularly Christianity, have assimilated the thesis as well as the antithesis, the former a part of the doctrine of the Original Sin, the latter of that of Divine Grace. Has science anything relevant to contribute to these ancient human insights? It can at least place them in the context of some well authenticated discoveries, which suggest that man is not predestined by his 'nature' to be good or bad, but may become either, depending on circumstances.

Ill. 206. Only man has free will and consequently the power to exercise egotism, altruism or heroism. Here a Buddhist monk, having been soaked in petrol, sets fire to himself and burns to death in front of thousands of onlookers, sacrificing his life for his political principles.

The pertinent evidence can be summarized under three rubrics. First and most evident is that the genetic inheritance, or 'nature', or innate tendencies, call them what you will, are only potentialities at the time when human life is kindled in the fertilized egg cell. The potentialities become realized in the process of individual development, in which the environment plays a part. Which potentialities are realized, and to what degree, depends on how a particular heredity interacts with particular environments. Secondly, man has two heredities, instead of a single one as in other organisms. His genetic heredity is coded in his genes and chromosomes. The cultural heredity, or simply culture, is transmitted not through genes but by precept, instruction, imitation, and learning. Mankind is the creator, and an individual person a creature of his culture. Thirdly, the capacity to be enculturated has been promoted in human evolution by natural selection. It has been promoted continuously and unrelentingly for two million years or longer. Why has natural selection furthered and maintained man's educability, his aptitude to profit by experience, and to adjust his behaviour to circumstances? The reason is that man's capacity to accept the culture of his society was and is indispensable for man's survival as an individual and as a species.

To forestall misunderstanding, while culture is acquired by learning and is not transmitted genetically, the capacity to acquire a culture is vouchsafed by man's biological inheritance. This capacity is an integral part of the genetic patrimony of the human species. Man is not preordained to develop into an altruist or an egotist, a saint or a sinner, hero or coward, hard worker or sluggard. He has potentialities to become any of the above kinds of persons. Human adaptedness is based on a remarkable open-endedness of the behaviour-development pattern. These statements are in need of a qualification. The developmental plasticity of human cultural behaviour is great but not infinite. People are not born identical *tabulae rasae*. Pathological conditions excepted, people can be trained for many kinds of work and occupations. Yet nobody is equally easily trainable for all conceivable callings. A remarkable freedom of choice of careers coexists in the human species with a genetic diversity of aptitudes. Some individuals are more easily trainable for some, other people for other occupations.

Ill. 207. The Chinese sage, Mencius, 371–289 BC.

Everyday experience clearly shows that people are endlessly diversified, not only in physical appearance but also in behaviour and attitudes. Human diversity may conceivably derive from upbringing and circumstances; it may also be innate. These explanations, obviously not mutually exclusive, have been under debate since antiquity. Aristotle declared that 'those who are sprung from better ancestors are likely to be better men, for nobility is excellence of family lineage'. Cicero, notwithstanding his having been a slave-owner, was convinced that all humans were basically alike. Eighteen and a half centuries later Jefferson, another slave-owner, held it to be self-evident truth that all men are born equal. Regardless of what philosophers thought about the causes of human diversity, social stratification and socio-economic inequality became established in all human societies.

Different people were deemed to exist for different purposes. The Chinese sage Mencius stated this quite unambiguously around 300 BC: 'Some labour with their minds and some labour with their physical strength. Those who labour with their minds rule others, and those who labour with their strength are ruled by others. Those who are ruled sustain others, and those who rule are sustained by others. This is a principle universally recognized.' The caste system in India was probably the most consistent application of this principle in a society, and it endured for nearly three millennia. Rigid social classes of feudal Europe were, in practice, not very different from castes. Their vestiges still survive. Yet the principle of human equality was also proclaimed, and eventually began to gain ascendance.

Human equality is not a biological phenomenon. Misinterpreting it as such is the source of much confusion. It is an ethical precept which is accepted, at least in theory, in some societies and not in others. It is ultimately religious in origin. In the Judaeo-Christian world view all humans, no matter how diverse individually, are God's children. They ought to seek the Kingdom of Heaven, although some will gain access thereto more easily than others. Jesus promised to the poor that the Kingdom of Heaven will be theirs, while the rich will have difficulties in getting there. Many who are among the first will be among the last, and the last will be first. Although the reversal of the status is to occur not in the here and now but in the hereafter, the idea is plainly subversive to the existent social orders. It has been systematically soft-pedalled by the established churches. Yet it has surfaced again and again in the course of history. Albigenses and Cathari, in the twelfth century, scorned affluence and social status, and became so alarming to the hierarchy that a crusade was organized to slaughter them. But their offshoot, the Waldenses, endured in some alpine valleys in Italy even to modern times. The Levellers, in seventeenth-century England, were an analogous phenomenon. The rise of human equality, intermittent, incomplete, and with many reverses, is however a phenomenon of modern times.

Equality of opportunity and of social status may usefully be distinguished. According to Eckland,[16] 'whereas equality of status deals with how power, privilege, and prestige are distributed in a society, equality of opportunity deals with the process of status allocation or, in other words, the criteria by which people are selected to fill different roles in society'. With equality of opportunity, the social and economic standings are achieved according to an individual's ability, and effort exerted to translate it into fulfilment. Under inequality of opportunity, social positions are ascribed on the basis of race, family prestige, etc. The concept of status has several dimensions; economic status is one of them; less tangible but perhaps no less important is the 'deference-position', in other words, the evaluation of the degree of prestige attached by members of a society to a given role.

Equality of opportunity and equality of status do not necessarily go together. Suppose that everybody can, regardless of race and family antecedents, aspire and compete for better education, greater income, and for any occupation or profession desired. Suppose further that the rules of the competition are entirely 'fair'; to many people this means that the outcome depends only on the abilities of the competing individuals, and not on the status or prestige of their families, and that the element of chance or 'luck' is minimized. The result will be a meritocracy, in which certain positions or roles in the society are esteemed more than others, and their holders are rewarded with high incomes and prestige.[29] Now, meritocracy does not imply equality of status, it only means that the status is 'merited' according to the notions prevalent in a given society. There will almost inevitably be an elite, or elites, people in positions of eminence, prestige, and high material rewards, as well as commoners doing necessary but humdrum drudgery, and perhaps paupers, failures and outcasts in the competition process. This is the same situation which existed without equality of opportunity, where the elites, commoners and outcasts were not selected according to their abilities, but rather inherited the status of their parents. Furthermore, as pointed out particularly by Scarr-Salapatek[42] and myself,[14] the elites under equality of opportunity are likely to be at least to some extent also genetic elites. While the alleged nobility of the old aristocratic 'blood' was largely mythical, with equality of opportunity such claims may acquire some substance.

In history, it was mostly the equality of opportunity that was fought for or against. And it is in the realm of equality of opportunity that appreciable success ('progress', if such equality is your ideal) has been achieved. There were times when it was believed that some people are by nature slaves and others masters; this belief is now rather uncommon. It was assumed that some people are born to command and others to obey, some belong to an elite entitled by their patrician birth to affluence and deference, while others must put up with poverty; this is still given faith by many, chiefly on the affluent side, but the credibility of the assumption is dwindling. Those belonging to the allegedly inferior races or classes now refuse 'to know their place', and aspire to share the fruits of the earth with their 'superiors' on a more nearly equal basis. This is sometimes labelled 'revolt of the masses' or 'revolution of raising expectations'.

The advance towards equality of status has been, considering the world as a whole, much less impressive than with equality of opportunity. Not infrequently people who are ardent partisans of the former are lukewarm towards the latter, or regard it as utopian and even undesirable. It seems to them only 'natural' that competence should be rewarded with high socio-economic standing, and incompetence scourged with low status. The allegedly classless societies of the U.S.S.R. and Eastern Europe have far from reached complete equality of opportunity, and they have status inequalities at present which, if anything, are greater than their bourgeois capitalist predecessors. A classless society is hardly conceivable without equality both of opportunity and of status. In 'a Christian perspective' Needham[37] sees that 'one of the main characteristics of the future world co-operative commonwealth of justice and comradeship is bound to be a classless society', and he believes that the most earnest or even 'heroic' efforts to produce a really classless society are now being made in China. However, Needham admits that China still has far to go towards this goal, and that 'recrudescence of private interest' is a danger to be guarded against. Also in a Christian perspective, Teilhard de Chardin[48] sees that 'it is mankind as a whole, collective humanity, which is called upon to perform the definitive act whereby the total force of terrestrial evolution will be released and flourish; an act in which the full consciousness of each individual man will be sustained by that of every other man, not only living but the dead'. This is, then, what people are for.

GENETIC IMPROVEMENT OF MANKIND

Insistence on human equality does not at all mean that mankind as it exists now is perfect. Improvement is not only imaginable but feasible, both culturally and biologically. For otherwise the evolution would have achieved its summit, and one could only wish that it should terminate, lest it produce deterioration. The hope for improvement never left mankind, even in the darkest periods of its long history. To quote Teilhard de Chardin again:[48] 'For a man who sees nothing at the end of the world, nothing higher than himself, daily life can only be filled with pettiness and boredom.' Being able to imagine an improved mankind is another unique faculty of the human species. Surely no other biological species has such a capacity. Countless sociological and biological utopias (and some anti-utopias) have been designed, and it is not my intention to add another. However, a discussion of the import of evolutionism on man's self-image would be missing too essential a component if it failed to mention the potentiality of improvement. Perspectives of human equality have been dwelt on above; biological improvement will now be very briefly considered.

The idea that mankind's genetic endowment can be improved by selective breeding is very old, going back at least to Plato's 'Republic'. In 1883, Francis Galton proposed the name eugenics for the 'study of agencies under social control, that may improve or impair the racial qualities of future generations, either physically or mentally'. During the first third of the twentieth century, the eugenical movement in Europe and America had the misfortune of being a captive of racists, political reactionaries, and finally of Nazi evildoers. Its reputation sank so low that heroic efforts of new generations of eugenists to cleanse eugenics of its unsavoury associations did not fully succeed. The former American Eugenics Society has recently become the Society for the Study of Human Biology. That the genetic endowment of mankind should be safeguarded against deterioration, and if possible improved upon, is irrefutable. Those who hold human genes to be something so sacred that they should be left alone, fail to realize that to do so would be a form of tampering with human genes that will have predictable and, in the long run, unfortunate consequences.

Burdens of genetic defects carried in human populations are heavy, and furthermore they will grow heavier as time goes on, unless measures are taken to counteract the trend. The best existing estimates are that about five per cent of infants are born with gross genetic malformations of various sorts. To this must be added that between twelve and fifteen per cent of pregnancies lasting more than 4-5 weeks end in spontaneous abortion, and two per cent result in stillbirths. Not exactly known, but a considerable proportion of these abortions and stillbirths are genetically caused. The above figures do not include genetic diseases and constitutional weaknesses that manifest themselves in adult persons. Nor do overall statistics give us a grasp of the colossal burden of human misery for which genetic defects are responsible. Early abortion brings disappointment to the prospective parents; how much stronger is the anguish of the parents whose child slowly but inexorably pines away and dies of a genetic disease; or consider genetic diseases that condemn a person to a lifelong misery from which only death is a release.

To be sure, not all genetic diseases are incurable. More and more therapies specific for this or that genetic disease are being discovered. A classical example is a diet nearly free of amino acid phenylalanine, which permits a child with phenylketonuria to develop nearly normally, or at least to avoid gross mental retardation. 'Euphenic' measures similar in principle to the 'cure' of phenylketonuria will hopefully become available for some genetic defects that are at present irremediable. Does this remove the necessity of controlling the genetic endowments of human populations? Alas, by no means. In the first place, euphenic treatments 'cure' the symptoms of genetic defects, not the defective genes. Hence, persons cured and enabled to reproduce may be parents of more individuals in need of cures in the following generations. Secondly, it is unlikely that reliable and fully efficient euphenic treatments will be available for every kind of genetic defect, especially for defects due to interactions of several genes (polygenic conditions). And thirdly, even if treatments for every genetic defect were available, the perspective of most or all infants born needing multiple treatments, each for his specific collection of defects, is a melancholy one. I am not one of those who like composing horrendous scenarios of the genetic decadence of mankind. However, prevention of the birth of individuals with serious genetic defects seems both ethically and pragmatically to be the most acceptable solution.

Techniques for controlling, or at least for lowering the incidence of genetic defects are rapidly becoming available. Of course, all of them can and hopefully will be improved in the future. Genetic counselling is the technique which provokes fewest doubts and objections. A genetic counsellor, who is usually a physician versed in human genetics, provides a person or a family with the best information available about the nature of a given genetic trait and its chances of being transmitted to the progeny. Counselling proffers advice, not an order. The counsellees can draw their own conclusions as to the desirability of their accepting the responsibilities of parenthood. If the risks of a genetic disease are considerable many people will avoid it. Of course, some people do not shrink from taking chances. Having been forewarned could still be useful, if the parents are psychologically and otherwise prepared to take the consequences.

Amniocentesis is a relatively new technique, which makes it possible to detect certain (but by no means all) genetic defects in the foetus long before birth. A small amount of the fluid surrounding the foetus is withdrawn, and the foetal cells are propagated in tissue culture to be studied cytologically and/or biochemically. If a serious genetic defect is diagnosed, the birth of the afflicted foetus can be prevented by timely abortion. Public opinion in recent decades has moved far towards acceptance of abortion when the prospective mother requests it; abortion of defective foetuses should meet less opposition than abortion for other motives. Examination by amniocentesis is laborious and expensive, and it is unlikely that it will be utilized for mass screening of pregnant women. It is utilized when there are valid reasons to suspect that a genetic defect may be present. For example, the risk of mongolism is much greater in children of women above forty years of age than in young women; mongolism is caused by the presence of an extra chromosome that can be seen under the microscope.

Genetic engineering or genetic surgery is in the realm of possibilities, not of techniques now available for use. It may some day become feasible to alter defective genes chemically in the direction of 'normal' or wholesome genes. Thus one could substitute genes needed for health in place of defective genes. This would be a 'cure' not of the symptoms alone but of the causative agents of genetic defects.

The techniques just considered belong to the class of 'negative eugenic' measures, in other words measures designed to do away with harmful, defective, disease-producing genes. Basically, negative eugenics may be considered a branch of medicine. It aims at the prevention or healing of a certain class of ailments or infirmities, which are clearly related to the hereditary endowment. Most people are likely to accept negative eugenic measures as they accept other medical treatments. Positive eugenics is a different matter, even though the boundary between negative and positive intervention is sometimes blurred. Positive eugenics seeks to improve the genetic endowment of mankind not only by elimination of harmful genes, but rather by encouraging the propagation of what is regarded a genetic elite, and eventually by making new genes for superior characteristics. Starting with Galton, eugenic propaganda tried to persuade people with 'superior' genes to have large families with many children.

More radical measures than requesting voluntary co-operation of some people to raise many children have been recommended. H. J. Muller, Julian Huxley and others urged 'germinal choice' or 'prenatal adoption'. Semen of superior men should be collected, preserved in deep frozen condition, and used to artificially inseminate as many women as possible. Techniques may some day be devised to collect not only sperms of superior men but also ova of superior women. Fertilization may take place, the fertilized ova left in artificial media to develop to the stage when implantation in the uterus must occur, and the young embryos transferred to the uteri of women who agree to bear them to birth. Cloning would involve implantation of nuclei of body cells of superior men or women into ova whose nuclei had been removed. Alternatively, cells from tissue cultures might be stimulated to develop like fertilized ova. The developing embryos are then to be implanted into the uteri of women who need not be their genetic mothers. Cloning might permit many individuals to have genetic endowments identical with that of the donor of the nuclei, and with each other. Genetic engineering may be developed to the point when artificial genes could be made, and introduced into human cells. Some such genes may be imagined to engender traits superior to any existing genes. Hence, the dream of 'New Man'.

The above all too brief review of eugenical techniques, now available and likely to be developed in the future, validates at least one crucial assumption. Human genetic endowments can be brought under control. The paramount difficulty of all eugenic programmes is, however, not biological but sociological and ethical. Can people agree on what constitutes an improvement of the human genetic base? Negative eugenics meets rather fewer difficulties than the positive one. Almost everybody will recognize that mankind would be better off without some genetic conditions. Who can possibly wish infants to be born with extra chromosomes causing mongolism? And this is regardless of whether or not some euphenic method might be developed that would alleviate the mongolism symptoms. There are hundreds of kinds of genetic defects that seem to be unconditionally deleterious. Their incidence in the human populations ought to be minimized, if they cannot be eliminated entirely.

Unanimity is, however, unlikely soon to be reached concerning genetic traits with which man of the future should be provided. Almost everybody has a vague image of human beings with desirable qualities, such as kindness, high intelligence, honesty, humaneness, propensity to love and to be loved. These qualities come, however, mostly from cultural rather than from biological heredity. Their heritability is probably low, if any exists at all. Moreover, who is to decide just what combination of these and other qualities, physical and mental, should the genetic improvement programme be designed to produce? Are we to aim at some sort of universal genius and paragon of all virtues, and make his genetic endowment the property of all future humanity? Or should one rather have more variety and some specialization, some people with superb development of a certain ability and other abilities in other persons? Suppose that the Muller-Huxley project of artificial insemination of many women by the sperm of superior men were to be adopted. Who are the superior males whose semen is to be collected and utilized? Is this to be decided by a committee of eminent biologists or will the decision be pre-empted by politicians? Assume, for the sake of argument, that an agreement is reached on what genetic endowment or endowments would be good for mankind to have in the twenty-first and twenty-second centuries. Who knows what shall be the optimum genetic base for a man of the thirtieth or fortieth centuries?

Many social scientists and some biologists believe that the genetic evolution of mankind has now been completed. Human genetic endowment will henceforward remain constant and uniform everywhere. Only cultural evolution is now going on. This is at best a half-truth, the more misleading because it sounds plausible. Mankind is the only biological species which evolves both genetically and culturally. It is incontrovertible that for many millennia man's adaptedness was promoted by cultural change, and this will continue to be so in the predictable future. Yet it is fallacious to think of genetic and cultural evolutions as being separate and independent. They are components of a cybernetic system bound together by mutual feedbacks. Culture has a genetic foundation. This genetic understructure must be protected from deterioration. This is what negative eugenics is all about. In one form or another it will have to be adopted. Positive eugenics is premature and may be left in abeyance for the time being. The idea of positive eugenics is logically sound, and it may bear fruit eventually. Hasty measures might however do more harm than good and, what is particularly hazardous, the harm may be irreversible.

ENVOY

What, then, is man's self-image in the light of the modern understanding of evolution? An admirable recent book of S. E. Luria is entitled *Life the Unfinished Experiment* (1973). The 'experiment' is 'unfinished' because life has not only evolved but continues to evolve. This is no less true of man— *Ill. 189* mankind is an unfinished experiment. Whose experiments are these? The adherents of religious traditions believe the universe to be God's enterprise. Yet man has discovered that the experiments are in progress, and that he is called upon to assist and collaborate in their conduct. An experiment, the outcome of which is predestined and is known in advance, is not worth making. Man does not however know the outcome. Within limits imposed by nature's laws, mankind can do its best to mould its future in accord with its ideas of the good and the beautiful.

Postscript

Vanne Goodall

Il est dangereux de trop faire voir à l'homme combien il est égal aux bêtes, sans lui montrer sa grandeur. Il est encore dangereux de lui trop faire voir sa grandeur sans sa bassesse. Il est encore plus dangereux de lui laisser ignorer l'un et l'autre. Mais il est très advantageux de lui representer l'un et l'autre.

Il ne faut pas que l'homme croie qu'il est égal aux bêtes, ni aux anges, ni qu'il ignore l'un et l'autre, mais qu'il sache l'un et l'autre.

L'homme n'est ni ange ni bête, et le malheur veut que qui veut faire l'ange fait la bête.

(Pascal, *Pensées*)

It is dangerous to show man how closely he resembles the beast without at the same time pointing out his greatness. It is also dangerous to stress his greatness without mention of his baseness. It is still more dangerous to let him ignore both. It is most important that he should be made aware of both.

Man must not believe that he is equal to beast or angel, nor must he ignore either, he must know both.

Man is neither angel nor beast, the pity is that if he tries to be angel he becomes beast. (*trans.* J.-C. Peissel)

Man has now entered a critical stage of his development, in which his survival is threatened by nuclear war, ecological disaster and the mounting pressures of a global existence. If he is to progress towards a more successful future for himself and his children, it is vital that he should try to arrive at a more profound understanding not only of his acts of cruelty, aggression and violence, but also of the source and strength of his motivations for survival and his capacity for love and tenderness, mercy and altruism. By doing so he may be able to cope more efficiently with the problems and dangers of the situation he has created in the world today and ultimately to transcend them.

The Quest for Man has been directed towards the attainment of this goal.

Bibliography

Ardrey, Robert. *African Genesis*. Reprint ed. Collins, London 1969; Atheneum,   GENERAL
New York 1961.

Beer, Gavin de. *A Handbook on Evolution*. British Museum (Natural History),
London 1970.

Boule, M., and Vallois, H. V. *Fossil Men*. Thames & Hudson, London 1957.

Buettner-Janusch, John. *The Origins of Man*. John Wiley, New York 1966.

Clark, Grahame. *The Stone Age Hunters*. Thames & Hudson, London 1967.

Clark, W. E. Le Gros. *The Antecedents of Man*. 2nd rev. ed. Edinburgh University
Press, Edinburgh 1971; Harper & Row, New York 1960.

—— *History of the Primates*. British Museum (Natural History), London 1962.

Cole, Sonia. *The Prehistory of East Africa*. Weidenfeld & Nicolson, London 1964.

—— *The Neolithic Revolution*. British Museum (Natural History), London 1970.

Coon, C. S. *The Origin of Races*. Jonathan Cape, London 1963.

Dart, R. A., and Craig, D. *Adventure with the Missing Link*. Hamish Hamilton,
London 1959.

Darwin, Charles. *The Origin of Species by Means of Natural Selection*. John Murray,
London 1859.

——*The Descent of Man, and Selection in Relation to Sex*. 2 vols. John Murray, London
1871.

Day, M. H. *A Guide to Fossil Man*. Cassell, London 1965.

Dobzhansky, T. *The Biology of Ultimate Concern*. New American Library, New York
1967; Rapp & Whiting, London 1969.

—— *Mankind Evolving*. Yale University Press, New Haven & London 1962.

Ehrlich, Paul R. *The Population Bomb*. Ballantine Books, New York 1968; Pan,
London 1971.

Hawkes, J., and Woolley, L. *Prehistory and the Beginnings of Civilization*. Allen &
Unwin, London 1963.

Howell, F. C. *Early Man*. Time-Life Books, New York 1965; Seymour Press,
London 1967.

Lawick-Goodall, J. van. *In the Shadow of Man*. Collins, London 1971; Houghton
Mifflin, Boston 1971.

Leakey, L. S. B. *Adam's Ancestors*. 4th ed. Methuen, London 1953; Harper & Row,
New York 1960.

Leakey, L. S. B., and Goodall, Vanne Morris. *Unveiling Man's Origins*. Schenkman,
Cambridge, Mass., 1969; Methuen, London 1970.

Lucretius Carus, Titus. *On the Nature of the Universe*. Harmondsworth 1951.

Lyell, Charles. *Principles of Geology*. 12th ed. 2 vols. London 1875.

Morris, Desmond. *The Naked Ape*. Jonathan Cape, London 1967; McGraw Hill,
New York 1967.

Napier, John. *The Roots of Mankind*. Smithsonian Institution Press, Washington
1970; Allen & Unwin, London 1971.

Oakley, K. P. *Man the Tool-maker*. 5th ed. British Museum (Natural History),
London 1967.

Pfeiffer, John E. *The Emergence of Man*. Harper & Row, New York 1969; Thomas
Nelson, London 1970.

Pilbeam, David. *The Evolution of Man*. London and New York 1970.

Simpson, George Gaylord. *The Meaning of Evolution*. Yale University Press, New Haven 1949; Oxford University Press, London 1950.

Sinnott, E. W. *The Biology of the Spirit*. Victor Gollancz, London 1956.

Teilhard de Chardin. *The Phenomenon of Man*. Harper & Row, New York 1959; Collins, London 1965.

Toynbee, A. J. *An Historian's approach to Religion*. London 1956.

Washburn, S. L. *Social Life of Early Man*. Aldine, Chicago 1961; Methuen, London 1962.

Weiner, J. S. *The Piltdown Forgery*. Oxford University Press, London 1955.

Wendt, Herbert. *From Ape to Adam*. Thames & Hudson, London 1972.

REFERENCES

(The numbers preceding the entries correspond to references in each chapter.)

Chapter 2

1 Birdsell, J. B. *Human Evolution*. Rand McNally, Chicago 1972.

2 Boulding, K. E. *The Meaning of the Twentieth Century*. Harper & Row, New York 1964; Allen & Unwin, London 1965.

3 —— 'The social system and the energy crisis'. *Science* 184 (1974): 255–7.

4 Bowlby, J. *Attachment and Loss*. Vol. I: *Attachment*. Hogarth Press and the Institute of Psycho-Analysis, London 1969.

5 —— *Attachment and Loss*. Vol. II: *Separation: Anxiety and Anger*. Hogarth Press and the Institute of Psycho-Analysis, London 1969.

6 Brown, H. Introduction: 'The growth and distribution of human population'. In *Are our Descendants Doomed?*, edited by H. Brown and E. Hutchings Jr., pp. 3–12. Viking Press, New York 1972.

7 Davis, K. 'The changing balance of births and deaths'. In *Are our Descendants Doomed?*, edited by H. Brown and E. Hutchings Jr., pp. 13–33. Viking Press, New York 1972.

8 Deutsch, M. *The Resolution of Conflict*. Yale University Press, New Haven and London 1973.

9 Dobzhansky, T. *Mankind Evolving*. Yale University Press, New Haven and London 1962.

10 Douglas, A. *Industrial Peacemaking*. Columbia University Press, New York and London 1962.

11 Ehrlich, P. R. and A. H. *Population, Resources, Environment*. 2nd ed. W. H. Freeman, San Francisco 1972.

12 Ehrlich, P. R. and A. H., and Holdren, J. P. *Human Ecology*. W. H. Freeman, San Francisco 1973.

13 Frank, J. D. *Sanity and Survival*. Random House, New York 1967; Barrie & Rockliff, The Cresset Press, London 1968.

14 Goldschmidt, W. *Man's Way*. Holt, Rinehart & Winston, New York 1959.

15 Hamburg, D. 'Relevance of recent evolutionary changes to human stress biology'. In *Social Life of Early Man*, edited by S. Washburn, pp. 278–88. Aldine, Chicago 1961; Methuen, London 1962.

16 —— 'Emotions in perspective of human evolution'. In *Expression of the Emotions of Man*, edited by P. H. Knapp, pp. 300–17. International University Press, New York 1963.

17 —— 'Evolution of emotional responses: Evidence from recent research on nonhuman primates'. In *Animal and Human*, edited by J. Masserman, pp. 39–52. Grune and Stratton, New York 1968.

18 —— 'Aggressive behavior of chimpanzees and baboons in natural habitats'. *J. Psychiat. Res.* 8 (1971): 385–98.

19 —— 'Crowding, stranger contact, and aggressive behavior'. In *Society, Stress and Disease*, edited by L. Levi, vol. I, pp. 209–18. Oxford University Press, London 1971.

20 Hamburg, D. A., Coelho, G. V., and Adams, J. E. 'Coping and adaptation: Steps toward a synthesis of biological and social perspectives'. In *Coping and Adaptation*, edited by G. V. Coelho, D. A. Hamburg and J. E. Adams. Basic Books, New York 1974.

21 Hamburg, D. A., and Lawick-Goodall, J. van. 'Factors facilitating development of aggressive behavior in chimpanzees and humans'. In *Determinants and Origins of Aggressive Behavior*, Mouton Publishers, The Hague 1974.

22 Hamburg, D. A., Adams, J. E., and Brodie, H. K. H. 'Coping behavior in stressful circumstances: Some implications for social psychiatry'. In *Further Explorations in Social Psychiatry*, edited by A. H. Leighton. Basic Books, New York 1974.

23 Hamburg, D. A. and B. A., and Barchas, J. 'Anger and depression in perspective of behavioural biology'. In *Parameters of Emotion*, edited by L. Levi. In Press. Raven Press, New York.

24 Hinde, R. A., and Davis, L. 'Changes in mother-infant relationship after separation in rhesus monkeys'. *Nature* 239 (1972): 41–2.

25 Holmes, T. H., and Rahe, R. H. 'The social readjustment rating scale'. *J. Psychosom. Res.* 11 (1967): 213–18.

26 Jeffrey, W. E. 'Perception, attention and curiosity'. In *Perspectives in Child Psychology*, edited by T. D. Spencer and N. Kass. McGraw-Hill, New York 1970.

27 Klerman, G. L. 'Depression and adaptation'. Presented at NIMH Conference on Psychology and Depression, Warrington, Virginia, 8–9 October 1971.

28 Lawick-Goodall, J. van. 'The behavior of free-living chimpanzees in the Gombe Stream area'. *Animal Behaviour Monographs* I (3) (1968): 161–311.

29 —— *In the Shadow of Man*. Collins, London 1971; Houghton Mifflin, Boston 1971.

30 —— 'Some aspects of aggressive behaviour in a group of free-living chimpanzees'. *Int. Soc. Sci. Journ.* 23 (1971): 89–97.

31 —— 'The behavior of chimpanzees in their natural habitat'. *American Journal of Psychiatry* 130 (1973): 1–12.

32 Lee, R. B., and DeVore, I. *Man the Hunter*. Aldine, Chicago 1968.

33 Lenski, G. *Human Societies*. McGraw-Hill, New York 1970.

34 LeVine, R. A., and Campbell, D. T. *Ethnocentrism: Theories of Conflict, Ethnic Attitudes and Group Behavior*. John Wiley, New York and London 1972.

35 Lindemann, E. 'Symptomatology and management of acute grief'. *American Journal of Psychiatry* 101 (1944): 141.

36 Lowe, J. R. *Cities in a Race with Time*. Random House, New York 1967.

37 MacLean, P. D. 'The triune brain, emotion and scientific bias'. In *The Neurosciences: Second Study Program*, edited by F. O. Schmitt, pp. 336–49. Rockefeller University Press, New York 1970.

38 Mumford, L. *The City in History*. Harcourt, Brace & World, New York 1961; Secker & Warburg, London 1961.

39 Napier, John. *The Roots of Mankind*. Smithsonian Institution Press, Washington 1970; Allen & Unwin, London 1971.

40 Parkes, C. M. *Bereavement: Studies of Grief in Adult Life*. International University Press, New York 1972; Tavistock Publications, London 1972.

41 Revelle, R. 'Some consequences of rapid population growth'. In *Are Our Descendants Doomed?*, edited by H. Brown and E. Hutchings Jr., pp. 42–64. Viking Press, New York 1972.

42 Sjoberg, G. *The Preindustrial City*. Free Press, New York 1960.

43 Smith, C. G. *Conflict Resolution: Contributions of the Behavioral Sciences*. University of Notre Dame Press, Notre Dame, Indiana, and London 1972.

44 Washburn, S. L., and Dolhinow, P. *Perspectives on Human Evolution*. Vol. 2. Holt, Rinehart & Winston, New York 1972.

45 Washburn, S. L., and McCown, E. R. 'Evolution of human behavior'. *Social Biology* 19 (1972): 163–70.

46 Washburn, S. L., and Moore, R. *Ape into Man*. Little, Brown & Co., Boston 1974.

47 White, B. *Human Infants*. Prentice-Hall, Englewood Cliffs, N. J., 1971.

48 Yarrow, L. J., and Pederson, F. A. 'Attachment: Its origins and course'. In *The Young Child, Reviews of Research*, edited by W. W. Hartup, vol. 2. National Association for the Education of Young Children, Washington, D.C., 1972.

49 Young, J. Z. *An Introduction to the Study of Man*. Oxford University Press, London 1971.

Chapters 3 & 4

1 Bada, J. L., and Protsch, R. 'Reacimization reaction of aspartic acid use in dating fossil bones'. *Proceedings of the National Academy of Sciences* 70 (1973): 1331–4.
2 Brace, C. L. 'The fate of the Classic Neanderthals'. *Current Anthropology* 5 (1964): 3–43.
3 Brain, C. K. 'New finds at the Swartkrans Australopithecine Site'. *Nature* 225 (1970): 1112–19.
4 Cartmill, Matt. In *Functional and Evolutionary Biology of Primates*, edited by Russell Tuttle. Aldine-Atherton, Chicago and New York 1972.
5 Day, Michael H. *A Guide to Fossil Man*. Cassell, London 1965.
6 —— 'The Omo human skeletal remains'. *Nature* 222 (1972): 1135.
7 Howells, William. *Evolution of the Genus Homo*. Addison-Wesley, Reading, Mass., 1973.
8 Isaac, Glynn Ll, and Curtis, Garniss H. 'Age of early Acheulian industries from the Peninj Group, Tanzania'. *Nature* 249 (1974): 624–7.
9 Jolly, C. J. 'The seed eaters: a new model of hominid differentiation based on a baboon analogy'. *Man* 5 (1970): 5–26.
10 Leakey, L. S. B., and Goodall, Vanne Morris. *Unveiling Man's Origins*. Schenkman, Cambridge, Mass., 1969; Methuen, London 1970.
11 Leakey, L. S. B., Tobias, P. V., and Napier, J. R. 'A new species of the genus *Homo* from Olduvai Gorge'. *Nature* 202 (1964): 5–7.
12 Leakey, M. D. 'Early artefacts from the Koobi Fora Area'. *Nature* 226 (1970): 228–30.
13 Leakey, R. E. F. 'Evidence for an advanced Plio-Pleistocene hominid from East Rudolf, Kenya'. *Nature* 242 (1973): 447–50.
14 Mayr, E. 'Taxonomic categories in fossil hominids'. *Cold Spring Harbor Symp. Quant. Biol.* 15 (1950): 109–18.
15 Napier, John. 'Locomotor Functions of Hominids'. In *Classification and Human Evolution*, edited by S. L. Washburn. Aldine, Chicago 1963.
16 —— 'Antiquity of Human Walking'. *Scientific American* 216 (1967): 56–66.
17 —— In *Old World Monkeys*, edited by J. R. and P. H. Napier. Academic Press, New York 1970.
18 —— *The Roots of Mankind*. Smithsonian Institution Press, Washington 1970; Allen & Unwin, London 1971.
19 Oakley, Kenneth P. *Man the Tool-maker*. 5th ed. British Museum (Natural History), London 1967.
20 Partridge, T. C. *Nature* 246 (1973): 75.
21 Pfeiffer, John E. *The Emergence of Man*. Harper & Row, New York 1969; Thomas Nelson, London 1970.
22 Pilbeam, David R. *The Ascent of Man*. Macmillan, New York 1972.
23 Simons, Elwyn L. *Primate Evolution*. Macmillan, New York 1972.
24 Simons, Elwyn L., and Pilbeam, David R. 'Preliminary revision of the Dryopithecinae'. *Folia primat.* 3 (1965): 81.
25 Tobias, P. V. In *Functional and Evolutionary Biology of Primates*, edited by Russell Tuttle. Aldine-Atherton, Chicago 1972.
26 Weiner, J. S., and Campbell, B. G. 'The taxonomic status of the Swanscombe Skull'. *Royal Anthropological Institute, Occasional Paper* No. 20 (1964): 175–209.

Chapter 5

1 Adams, R. M. *The Evolution of Urban Society: Early Mesopotamia and Prehispanic Mexico* (p. 38). Weidenfeld & Nicolson, London 1966.
2 Bender, B. *Farming in Prehistory: from Hunter-gatherer to Food Producer*. In press.
3 Bordes, F. H. *The Old Stone Age*. Weidenfeld & Nicolson, London 1968.
4 Boserup, E. *The Conditions of Agricultural Growth*. London 1965.
5 Harlan, J. R. 'A wild wheat harvest in Turkey'. *Archaeology* 20 (1961): 197.

6 Hole, F., Flannery, K. V., and Neely, J. A. *Prehistory and Human Ecology of the Deh Luran Plain*. Ann Arbor 1969.

7 Howell, F. C. *Early Man* (pp. 85, 128). Time-Life Books, New York 1965; Seymour Press, London 1967.

8 Kenyon, K. M. 'The origins of the Neolithic'. *The Advancement of Science* 26 (1969): 144.

9 Kirkbride, D. 'Five seasons at the pre-pottery Neolithic village of Beidha in Jordan'. *Palestine Exploration Quarterly* 98 (1966): 8.

10 Lee, R. 'Work effort, group structure and land use in contemporary hunter-gatherers'. In *Man, Settlement and Urbanism*, edited by P. J. Ucko, R. Tringham and G. W. Dimbleby, p. 177. Duckworth, London 1972.

11 —— 'Population growth and the beginnings of sedentary life among the !Kung Bushmen'. In *Population Growth: Anthropological Implications*, edited by B. Spooner, p. 329. MIT Press, Cambridge, Mass., and London 1972.

12 MacNeish, R. S. 'A summary of the subsistence'. In *The Prehistory of the Tehuacán Valley*, edited by D. S. Byers, vol. I, p. 290. University of Texas Press, Austin and London 1967.

13 —— 'Conclusion'. In *The Prehistory of the Tehuacán Valley*, edited by D. S. Byers, vol. I, p. 227. University of Texas Press, Austin and London 1967.

14 Mellaart, J. *Çatal Hüyük. A Neolithic Town in Anatolia*. Thames & Hudson, London 1967.

15 Perrot, J. 'Eynan (Aïn Mallaha)'. *Revue Biblique* 69 (1962): 384.

16 Renfrew, C. *The Emergence of Civilization. The Cyclades and the Aegean in the Third Millennium BC* (p. 13). Methuen, London 1972.

17 Renfrew, J. *Palaeoethnobotony. Prehistoric Food Plants of the Near East and Europe*. Methuen, London 1973.

18 Sanders, W. T., and Price, B. J. *Meso-America. The Evolution of a Civilization* (p. 227). Random House, New York 1968.

19 Smith, C. E. 'Plant remains'. In *The Prehistory of the Tehuacán Valley*, edited by D. S. Byers, vol. I, p. 220. Austin and London 1967.

20 Srejović, D. *Europe's first Monumental Sculpture: New Discoveries at Lepenski Vir*. Thames & Hudson, London 1972.

21 Steward, J. H. 'Ethnography of the Owens Valley Paiute'. *University of California Publications in American Archaeology and Ethnology* 33 (1933): 233.

22 Ucko, P. J., and Rosenfeld, A. *Palaeolithic Cave Art*. Weidenfeld & Nicolson, London 1967.

23 Wolf, E. *Peasants* (p. 9). Prentice-Hall, Englewood Cliffs, N.J., 1966.

Chapter 6

1 Beatty, H. 'A note on the behavior of the chimpanzee'. *Journal of Mammalogy* 32 (1951): 118.

2 Boelkins, R. C., and Wilson, A. P. 'Intergroup social dynamics of the Cayo Santiago rhesus (*Macaca mulatta*) with special reference to changes in group membership by males'. *Primates* 13, 2 (1972): 125–40.

3 Butler, R. A. 'Investigative behaviour'. In *Behaviour of Non-human Primates*, edited by A. M. Schrier, H. F. Harlow and F. Stollnitz. Academic Press, London and New York 1965.

4 Bygott, J. D. 'Cannibalism among wild chimpanzees'. *Nature* 238 (1972): 410–11.

5 DeVore, I., and Hall, K. R. L. 'Baboon ecology'. In *Primate Behaviour*, edited by I. DeVore. Holt, Rinehart & Winston, New York 1965.

6 DeVore, I., and Washburn, S. L. 'Baboon ecology and human evolution'. In *African Ecology and Anthropology* 36. New York 1963.

7 Eibl-Eibesfeldt, I. *Ethology, the Biology of Behaviour* (p. 279). Holt, Rinehart & Winston, New York 1970.

8 Fossey, Dian. Personal communication 1972.

9 Gallup, G. G. 'Chimpanzees: self-recognition'. *Science* 167 (1970): 86–7.

10 Gardner, R. A. and B. T. 'Teaching sign language to a chimpanzee'. *Science* 165 (1969): 664–72.

11 Halperin, S. Personal communication 1973.

12 Harding, R. S. O. 'Predation by a troop of olive baboons (*Papio anubis*)'. *Proceedings of Congress. Am. J. Phys. Anthr.* 38 (1973): 587–92.

13 Hayes, C. *The Ape in our House*. Harper & Row, New York 1951; Victor Gollancz, London 1952.

14 Kawabe, M. 'One observed case of hunting behaviour among wild chimpanzees living in the savannah woodland of western Tanzania'. *Primates* 7 (3) (1966): 393–6.

15 Kohler, W. *The Mentality of Apes*. Harcourt Brace, New York 1925.

16 Kohts, N. 'Infant ape and human child'. *Science Memorial Museum for Darwin*, Moscow 3 (1935): 1–586.

17 Kummer, H. *Social Organization of Hamadras Baboons*. University of Chicago Press, Chicago and London 1968.

18 Lawick-Goodall, J. van. 'The behaviour of free-living chimpanzees in the Gombe Stream area'. *Animal Behaviour Monographs* I (3) (1968): 161–311.

19 —— *In the Shadow of Man*. Collins, London 1971; Houghton Mifflin, Boston 1971.

20 —— 'Cultural elements in a chimpanzee community'. In *Pre-cultural Primate Behaviour*, edited by E. W. Menzel. Karger Publications 1973.

21 Lawick-Goodall, J. van, and Hamburg, D. A. 'New evidence on the origins of human behaviour'. In *American Handbook of Psychiatry*, edited by D. A. Hamburg and K. Brodie. Vol. 6: *New Frontiers*. Basic Books, New York 1974.

22 Lawick-Goodall, J. van, and Lawick, H. van. *Innocent Killers*. Collins, London 1970; Houghton Mifflin, Boston 1970.

23 McGinnis, P. R. 'Patterns of sexual behaviour in a community of free-living chimpanzees'. Unpublished doctoral dissertation for Cambridge University, 1973.

24 McGrew, W. C., and Tutin, C. E. G. 'Chimpanzee Dentistry'. *Journal of American Dental Association* 85 (1972): 1198–204.

25 Menzel, E. W. 'Patterns of responsiveness in chimpanzees reared through infancy under conditions of environmental restriction'. *Psychologische Forschung* 27 (1964): 337–65.

26 Morris, Desmond. *The Naked Ape*. Jonathan Cape, London 1967; McGraw Hill, New York 1967.

27 Nissen, H. W. 'A field study of the chimpanzee'. *Comparative Psychology Monographs* 8 (1). John Hopkins Press (Pioneering study), Baltimore 1931.

28 Premack, D. 'Language in Chimpanzee?'. *Science* 172 (1971): 808–22.

29 Pusey, A. Personal communication 1972.

30 Rensche, B., and Döhl, J. 'Wahlen Zwischen zwei überschaubaren Labyrinthwegen durch einen Schimpansen'. *Z.f. Tierpsychol.* 25 (1968): 216–31.

31 Sade, D. S. 'A longitudinal study of social behaviour of rhesus monkeys'. In *The Functional and Evolutionary Biology of Primates*, edited by Russell Tuttle. Aldine-Atherton, Chicago 1972.

32 Savage, T. S., and Wyman, J. *Boston Journal of Natural History* 4 (1843–44).

33 Sugiyama, Y. 'Social behaviour of chimpanzees in the Budongo Forest, Uganda'. *Primates* 10 (1969): 197–225.

34 Suzuki, A. 'Carnivority and cannibalism observed among forest-living chimpanzees'. *Journal of Anthropological Society, Nippon* 79 (1) (1971): 30–48.

35 Teleki, G. 'The omnivorous chimpanzee'. *Scientific American* 228 (1973).

36 Thorndahl, M. Personal communication 1973.

37 Tutin, C. E. G. Personal communication 1974.

38 Washburn, S. L., and DeVore, I. 'Social behaviour of baboons and early man'. In *Social Life of Early Man*, edited by S. L. Washburn. Aldine, Chicago 1961; Methuen, London 1962.

39 Yerkes, R. M. *Chimpanzees: A Laboratory Colony*. Yale University Press, New Haven 1945.

Chapter 7

1 Benedict, R. F. *Patterns of Culture*. Houghton Mifflin, Boston 1934; Routledge & Kegan Paul, London 1961.

2 Bilz, R. 'Zur Grundlegung einer Paläopsychologie: I. Paläophysiologie. II. Paläopsychologie'. *Schweiz. Z. Psychol.* 3 (1944): 202–12, 272–80.

3 Boas, F. *The Social Organization and the Secret Societies of the Kwakiutl Indians*, 1897. Reprint, Johnson's Reprint Corp., New York 1970.

4 Brownlee, F. 'The Social Organization of the Kung (!Un) Bushmen of the North-Western Kalahari'. *Africa* 14 (1943): 124–9.

5 Eibl-Eibesfeldt, I. *Die !Ko-Buschmanngesellschaft. Gruppenbindung und Aggressionskontrolle.* Monographien zur Humanethologie I. Piper, Munich 1972.

6 —— *Der Vorprogrammierte Mensch. Das Ererbte als bestimmender Faktor im menschlichen Verhalten.* Molden, Vienna 1973.

7 Heinz, H. J. 'The Social Organization of the !Ko Bushmen'. Master's thesis, Dept. of Anthropology, University of South Africa, Johannesburg 1966.

8 —— 'Conflicts, Tensions and Release of Tensions in a Bushman Society'. The Institute for the Study of Man in Africa, Isma Papers No. 23 (1967).

9 —— 'Territoriality among the Bushmen in general and the !Ko in particular'. *Anthropos* 67 (1972): 405–16.

10 Helmuth, H. 'Zum Verhalten des Menschen: Die Aggression'. *Z. Ethnol.* 92 (1967): 265–73.

11 König, H. 'Der Rechtsbruch und sein Ausgleich bei den Eskimo'. *Anthropos* 20 (1925): 276–315.

12 Kohl-Larsen, L. *Wildbeuter in Ostafrika. Die Tindiga, ein Jäger- und Sammlervolk.* Reimer, Berlin 1958.

13 Lawick-Goodall, J. van. *In the Shadow of Man.* Collins, London 1971; Houghton Mifflin, Boston 1971.

14 Lebzelter, V. *Eingeborenenkulturen von Süd- und Südwestafrika.* Leipzig 1934.

15 Lee, R. B. 'What Hunters do for a Living'. In *Man the Hunter*, edited by R. B. Lee and I. DeVore. Aldine, Chicago 1968.

16 —— 'The !Kung Bushmen of Botswana'. In *Hunters and Gatherers Today*, edited by M. G. Bicchieri, pp. 327–68. Holt, Rinehart & Winston, New York 1973.

17 Lee, R. B., and DeVore, I. *Man the Hunter.* Aldine, Chicago 1968.

18 Marshall, L. 'The !Kung Bushmen of the Kalahari Desert'. In *Peoples of Africa*, edited by J. L. Gibbs, pp. 241–78. New York 1965.

19 Passarge, S. *Die Buschmänner der Kalahari.* D. Reimer, Berlin 1907.

20 Ploog, D. 'Verhaltensforschung und Psychiatrie'. In *Psychiatrie der Gegenwart*, I, IB, edited by H. W. Gruhle, R. Jung, W. Mayer-Gross and M. Müller, pp. 291–443. Springer, Berlin 1964.

21 Rasmussen, K. *People of the Polar North.* Kegan Paul, London 1908.

22 Rothmann, M., and Teuber, E. 'Einzelausgabe der Anthropoidenstation auf Teneriffa: I. Ziele und Aufgaben der Station sowie erste Beobachtungen an den auf ihr gehaltenen Schimpansen'. *Abh. Preuss. Akad. Wiss.*, Berlin (1915): 1–20.

23 Sahlins, M. D. 'The Origin of Society'. *Scientific American* 204 (1960): 76–87.

24 Sbrzesny, H. '!Ko-Buschleute (Kalahari)—Der Eland-Tanz. Kinder spielen das Mädchen-Initiationsritual'. *Homo* 24 (4) (1974).

25 Schebesta, P. 'Die Bambuti-Pygmäen vom Ituri'. *Mem. Institut Royal Colonial Belge* 2 (1) (1941): 1–284.

26 —— 'Die Bambuti Pygmäen vom Ituri'. *Mem. Institut Royal Colonial Belge* 2 (2) (1948): 285–551.

27 Schjelderup, H. *Einführung in die Psychologie.* Huber, Berne 1963.

28 Schmidbauer, W. *Jäger und Sammler.* Selecta, Planegg vor Munich 1972.

29 Silberbauer, G. B. 'Socio-Ecology of the G/wi Bushmen'. Thesis, Department of Anthropology and Sociology, Monash University, Australia 1973.

30 —— 'The G/wi Bushmen'. In *Hunters and Gatherers Today*, edited by M. G. Bicchieri, pp. 271–326. Holt, Rinehart & Winston, New York 1973.

31 Stow, G. W., and Bleek, D. F. *Rock-Paintings in South Africa.* London 1930.

32 Tobias, P. V. 'Bushmen—Hunter-gatherers. A Study in Human Ecology'. In *Ecological Studies in Southern Africa*, edited by D. H. S. Davis. W. Junk, The Hague 1964. Reproduced in *Man in Adaptation*, edited by Y. A. Cohen. Aldine, Chicago 1968.

33 Traill, A. 'The Complete Guide to the Koon'. Research Report in Linguistic Fieldwork undertaken in Botswana and South-west Africa, July 1972 and January 1973.

34 Vedder, H. 'Die Buschmänner Südwestafrikas und ihre Weltanschauung'. *South African Journal of Science* 34 (1937): 416–36.

35 —— 'Uber die Vorgeschichte der Völkerschaften von Südwestafrika'. *Journal from South West Africa Sc. Soc.* 9 (1952–3): 45–56.

36 Wilhelm, J. H. 'Die !Kung Buschleute'. *Jahrb. d. Museums f. Völkerkunde Leipzig* 12 (1953): 91–189.

37 Wickler, W. 'Socio-sexual signals and their intraspecific imitation among primates'. In *Primate Ethology*, edited by D. Morris, pp. 69–147. London 1967.

38 Woodburn, J. 'Stability and flexibility in Hadza residential groupings'. In *Man the Hunter*, edited by R. B. Lee and I. DeVore, pp. 103–110. Aldine, Chicago 1968.

39 Zastrow, B. V., and Vedder, H. 'Die Buschmänner'. In *Das Eingeborenenrecht: Togo, Kamerun, Südwestafrika, die Südseekolonien*, edited by E. Schultz-Ewerth and L. Adam. Strecker & Schröder, Stuttgart 1930.

Chapter 8

1 Adler, M. *The Difference of Man and the Difference it Makes.* Holt, Rinehart & Winston, New York 1967.

2 Ardrey, R. *African Genesis.* Reprint ed. Collins, London 1969; Atheneum, New York 1961.

3 Bidney, D. *Theoretical Anthropology.* New York and London 1953.

4 Birch, C. L. 'Participatory evolution; the drive of creation'. *Journal Amer. Acad. Religion* 90 (1972): 147–163.

5 Campbell, B. G. *Human Evolution.* Aldine, Chicago 1966; Heinemann, London 1967.

6 Darwin, Charles. *The Origin of Species by Means of Natural Selection.* John Murray, London 1859.

7 —— *The Descent of Man, and Selection in Relation to Sex.* 2 vols. John Murray, London 1871.

8 DeVore, I. (ed.) *Primate Behaviour: Field Studies on Monkeys and Apes.* Holt, Rinehart & Winston, Boston and London 1965.

9 Dobzhansky, T. *Evolution, Genetics, and Man.* John Wiley, New York 1955; Chapman & Hall, London 1955.

10 —— *Mankind Evolving.* Yale University Press, New Haven and London 1962.

11 —— *The Biology of Ultimate Concern.* New American Library, New York 1967; Rapp & Whiting, London 1969.

12 —— 'Darwin versus Copernicus'. In *Changing Perspective on Man*, edited by B. Rothblatt. Chicago University Press, Chicago and London 1968. Also in: Dobzhansky, T. *Genetic Diversity and Human Equality.* Basic Books, New York 1973.

13 —— *Genetics of the Evolutionary Process.* Columbia University Press, New York and London 1970.

14 —— 'Is genetic diversity compatible with human equality?' *Soc. Biol.* 20 (1973): 280–88.

15 —— 'Ethics and values in biological and cultural evolution'. *Zygon* 8 (1973) 261–81.

16 Eckland, B. K. 'Trends in the direction and intensity of natural selection, with special reference to the effects of destratification and equal opportunity upon current birth rates'. *American Eugenics Society Symposium on Human Evolution*, New York 1972.

17 Frisch, K. von. *The Dancing Bees.* Methuen, London 1954; Harcourt Brace, New York 1955.

18 Fromm, S. *The Heart of Man.* Harper & Row, New York 1964; Routledge & Kegan Paul, London 1965.

19 Gardner, R. A. and B. T. 'Teaching sign language to a chimpanzee'. *Science* 165 (1969): 664–72.

20   Grasse, P. P. *L'Evolution du Vivant*. Albin Michel, Paris 1973.
21   Greene, J. C. *The Death of Adam*. Iowa State University Press, Ames., 1959.
22   Hallowell, A. I. 'The protocultural foundations of human adaptation'. In *Social Life of Early Man*, edited by S. L. Washburn. Aldine, Chicago 1961; Methuen, London 1962.
23   Hamburg, D. A. 'Aggressive behavior of chimpanzees and baboons in natural habitats'. *J. Psychiat. Res.* 8 (1971): 385–98.
24   Hayes, C. *The Ape in our House*. Harper & Row, New York 1951; Victor Gollancz, London 1952.
25   Hayes, R. J. and C. 'The cultural capacity of chimpanzee'. *Human Biology* 26 (1954): 288–303.
26   Hofer, T., and Altner, G. *Die Sonderstellung des Menschen*. Stuttgart 1972.
27   Holloway, R. L. 'Culture, a human domain'. *Current Anthropology* 5 (1969): 135–68.
28   Hulse, F. *The Human Species*. Random House, New York 1971.
29   Jencks, C. *Inequality*. Basic Books, New York 1972.
30   Langer, S. R. 'The Lord of Creation'. *Fortune*, January 1944; reprinted in *The Borzoi College Reader*, edited by C. Muscatine and M. Griffith. A. Knopf, New York 1966.
31   Lawick-Goodall, J. van. 'The behavior of free-living chimpanzees in the Gombe Stream area'. Animal Behaviour Monographs 1 (3) (1968): 161–311.
32   —— *In The Shadow of Man*. Collins, London 1971; Houghton Mifflin, Boston 1971.
33   Lawick-Goodall, J. van, and Hamburg, D. A. 'Chimpanzee behavior as a model for the behavior of early man'. In *American Handbook of Psychiatry*, edited by D. A. Hamburg and H. K. Brodie, vol. 6. Basic Books, New York 1974.
34   Matson, I. W. *The Broken Image*, Braziller, New York 1964.
35   Monod, J. *Chance and Necessity*. A. Knopf, New York 1971; Collins, London 1972.
36   Morris, Desmond. *The Naked Ape*. Jonathan Cape, London 1967; McGraw-Hill, New York 1967.
37   Needham, J. 'Hope and social evolution: T'Hien Hsia Ta Tung and Regnum Dei. A Christian perspective on the Chinese experience.' *Anticipation*, No. 14. World Council of Churches, Geneva 1973.
38   Premack, D. 'Language in chimpanzees?' *Science* 172 (1971): 808–22.
39   Rensch, B. *Evolution above the species level*. Methuen, London 1959; Columbia University Press, New York 1960.
40   —— *Biophilosophy*. Columbia University Press, New York 1971.
41   Scarr-Salapatek, S. 'Unknowns in the IQ equation'. *Science* 174 (1971): 1223–8.
42   —— 'Race, Social class, and IQ'. *Science* 174 (1971): 1285–95.
43   Sherrington, C. *Man on his Nature*. 2nd ed. Cambridge University Press, London 1951; Doubleday Anchor, Garden City, New York 1953.
44   Simpson, G. G. *The Major Features of Evolution*. Columbia University Press, New York 1953.
45   —— *This View of Life*. Harcourt Brace, New York 1964.
46   —— *Biology and Man*. Harcourt Brace, New York 1969.
47   Skinner, B. F. *Beyond Freedom and Dignity*. A. Knopf, New York 1971; Jonathan Cape, London 1972.
48   Teilhard de Chardin, P. *The Phenomenon of Man*. Harper & Row, New York 1959; Collins, London 1965.
49   Washburn, S. L. (ed.) *Social Life of Early Man*. Aldine, Chicago 1961; Methuen, London 1962.
50   —— *Classification and Human Evolution*. Aldine, Chicago 1963; Methuen, London 1964.
51   Washburn, S. L., and Lancaster, C. S. *Perspectives on Human Evolution*, edited by S. L. Washburn and P. C. Jay. New York 1968.
52   White, L. *The Science of Culture*. Grove Press, New York 1949.
53   Whitehead, A. N. *Science and the Modern World*. New York 1925; London 1926.

List of Illustrations

The authors and publishers are grateful to the many official bodies, institutions and individuals mentioned below for their assistance in supplying illustrative material.

1 Michelangelo: *Creation of Adam.* From the ceiling of the Sistine Chapel, Vatican. Photo: Phaidon Archives.

2 Gold bull's head from a lyre found at Ur. British Museum, London. Photo: Hirmer Fotoarchiv, Munich.

3 A Babylonian lyre: detail of the Royal Standard of Ur, *c.* 2700 BC. Photo: by courtesy of the Trustees of the British Museum, London.

4 Clay tablet containing fragments of the epic of Gilgamesh, from Assurbanipal's clay tablet library. Photo: by courtesy of the Trustees of the British Museum, London.

5 Terracotta statue of Gilgamesh. Musée du Louvre, Paris. Photo: Service de Documentation Photographique de la Réunion des Musées Nationaux, Paris.

6 Black outline painting of a wounded bison, a 'man', and a bird in the Cave of the Dead Man at Lascaux, Dordogne. By courtesy of the Caisse Nationale des Monuments Historiques, Paris. Photo: Jean Vertut.

7 Boucicaut Workshop: *God Presenting Eve to Adam.* Ms. 251, f. 16, Fitzwilliam Museum, Cambridge. Photo: Phaidon Archives.

8 Mosaic from near Pompeii showing *Plato's Academy.* Before AD 79. Museo Nazionale, Naples. Photo: Scala.

9 Title page of a Renaissance edition of Lucretius' *De Rerum Natura,* Antwerp 1565. Photo: John Freeman.

10 Leonardo da Vinci: *The Proportions of the Human Figure,* after Vitruvius. Pen and ink, *c.* 1492. Academy, Venice. Photo: Phaidon Archives.

11 Jean Muller explaining the Ptolemaic System. Fascimile of a wood engraving from *Epitome . . . Johannis de Monte-Regio,* 1543. Photo: Mary Evans Picture Library.

12 Italian Ptolomaic map of the world, 15th century. Photo: by courtesy of the National Maritime Museum, London.

13 *Time.* Photo composition by Chris Yates.

14 Lower Palaeolithic flint hand-axe found by John Frere in 1797 at Hoxne, Suffolk. From *Archaeologia,* 1800.

15 Fossil skeleton discovered by Scheuchzer. From J. Scheuchzer, *Physica Sacra,* 1731. Photo: Imitor.

16 Georges, Baron Cuvier (1769–1832). From a commemorative medal. Photo: Ronan Picture Library.

17 Charles Lyell (1797–1875). From *The Illustrated London News,* 1865. Photo: Ronan Picture Library.

18 Charles Darwin (1809–1882), in 1878. Photo: Mansell Collection.

19, 20 Contemporary cartoons from *Punch, c.* 1861, in connection with Darwin's theory of evolution. Photos: Mansell Collection.

21 The launching of a Mariner spacecraft from Cape Kennedy on 3 November 1973. Photo: N.A.S.A.

22 Ruins of a Roman Catholic church, in Quang Tri, Vietnam. Photo: Lee Rudakewych, Camera Press Ltd.

23 Contrasting bridges on the Trabia motorway near Palermo, Sicily. Photo: Grazia Neri.

24 Time-scale of the evolution of the earth. Drawn by J.-C. Peissel.

25 'Barmitzvah' confirmation ceremony on Mount Zion. Photo: Peter Larsen.

26 A Sunday school in progress, St. Francis Church, South Croydon. Photo: Barnaby's Picture Library. © Cleland Rimmer.

27 Dürer woodcut of a printing press, 1520. Photo: Phaidon Archives.

28 Satellite view of the earth. Photo: N.A.S.A.

29 Fish-eye view of a city. © Marshall Cavendish Ltd. Photo: Clay Perry.

30 Doré: *Over London by Rail.* Photo: Mary Evans Picture Library.

31 A woman returns to her slum home, demolition of which has been deferred. Photo: Nick Hedges, Shelter Picture Library.

32 Holbein: *Thomas More's Family,* preliminary sketch, 1527. Photo: Phaidon Archives.

33 The grief-stricken faces of bereaved parents of earthquake victims. Photo: Camera Press Ltd.

34 Christmas shoppers in Petticoat Lane, London. Photo: Horst Kolo.

35 The affection between a mother and child: Karen Hayes and her daughter Hilary. Photo: Alma McGoldrick, Camera Press Ltd.

36 Pollution. Photo: Peter Larsen.

37 Demonstrators taking part in a rally against the 1967 Abortion Act, 29 April 1974. Photo: Keystone Press Agency Ltd.

leaving for the hunt. Photo: Irven DeVore/Anthro-Photo.

178 A rock-painting from Battle Cave, Basutoland, showing Bushgirls going to collect vegetables. Drawn by J.-C. Peissel after Willcox.

179 A Bushman rock-painting showing a cattle raid. From Karl Weule, *Der Krieg in den Tiefen der Menschheit*, Stuttgart 1916.

180 An !Kung Bushboy tries to scratch his younger brother. Photo: I. Eibl-Eibesfeldt.

181 An !Kung Bushboy crying with frustration as his mother prevents him from scratching his brother. Photo: I. Eibl-Eibesfeldt.

182 An !Ko Bushboy pushing over a little girl who has tried to take a toy away from him. Photo: I. Eibl-Eibesfeldt.

183 An !Ko Bushgirl about to throw a stick at a boy who has teased her. Photo: I. Eibl-Eibesfeldt.

184 A model of the DNA spiral. Crown Copyright. Science Museum, London.

185 An Arabic map of the world dating from AD 1154, from a manuscript written in Cairo in 1456. Ms. Pococke 375, ff. 3v–4r, Bodleian Library, Oxford. Photo: by courtesy of the Curators of the Bodleian Library.

186 Nicolaus Copernicus' chart of the solar system. British Museum. Photo: Fleming.

187 The Horse-Head Nebula, south of Orion. Photo: Science Museum, London.

188 The expulsion from the Garden of Eden, from an early illustrated version of the Septuagint. Photo: © Marshall Cavendish Ltd.

189 Photo-montage symbolizing the eternal quest of the human spirit. Photo: John Garrett.

190 Rembrandt: *Dr. Nicholaes Tulp Demonstrating the Anatomy of the Arm.* Photo: Phaidon Archives.

191 Drawings of the brain by Leonardo da Vinci, from one of his anatomical sketchbooks. Photo: Phaidon Archives.

192 Drawings of hands by Albrecht Dürer. Photo: Phaidon Archives.

193 A measured anatomical drawing by Albrecht Dürer. Photo: Phaidon Archives.

194 Worker honey bees licking 'queen substance' from the queen's body and feeding her. Bruce Coleman Ltd. Photo: Colin G. Butler.

195–197 Washoe, a female chimpanzee, uses sign language to construct a sentence. Photos: by courtesy of Dr. Roger S. Fouts.

198 Reconstruction, by Maurice Wilson, of Peking man at work in the mouth of a cave. Photo: by courtesy of the British Museum (Natural History).

199 Pebble tools found in the caves at Chou-k'ou-tien. Reproduced by courtesy of Dr. Kenneth P. Oakley.

200 Hans Baldung: *Death and the Maiden.* Photo: by courtesy of the Kunsthistorisches Museum, Vienna.

201 The Shanidar Cave in northern Iraq, where Professor Ralph Solecki discovered nine skeletons of Neanderthal man. Photo: by courtesy of Professor Ralph S. Solecki.

202 A Neanderthal skeleton, lying as it was found in the Shanidar Cave, buried with garlands of flowers. Photo: by courtesy of Professor Ralph S. Solecki.

203 Ivory head of a French Gravettian 'Venus' figurine from Brassempouy. Musée des Antiquités Nationales, Saint-Germain-en-Laye. Photo: Thames & Hudson.

204 A Victorian print entitled *The Last of the Line.* Photo: Mary Evans Picture Library.

205 Michelangelo: *The Temptation and Fall of Man.* From the ceiling of the Sistine Chapel, Vatican. Photo: Phaidon Archives.

206 A Buddhist monk burning to death in front of thousands of onlookers at a main highway intersection in Saigon, Vietnam, 11 June 1963, in protest against what he called 'government persecution of Buddhists'. Photo: Associated Press Ltd.

207 Drawing of the Chinese sage, Mencius, 371–289 BC. Photo: Mary Evans Picture Library.

Index

Numbers in *italic* refer to illustrations and their captions